Individual Differences in Speech Production and Perception

SPEECH PRODUCTION AND PERCEPTION

Edited by Susanne Fuchs and Pascal Perrier

VOLUME 3

Notes on the quality assurance and peer review of this publication:
Prior to publication, the quality of the work published in this series
is reviewed by external referees appointed by the editorship.

Susanne Fuchs / Daniel Pape /
Caterina Petrone / Pascal Perrier (eds.)

Individual Differences in Speech Production and Perception

PETER LANG
EDITION

Bibliographic Information published by the Deutsche Nationalbibliothek
The Deutsche Nationalbibliothek lists this publication in the Deutsche Nationalbibliografie; detailed bibliographic data is available in the internet at http://dnb.d-nb.de.

Library of Congress Cataloging-in-Publication Data
Individual differences in speech production and perception / Susanne Fuchs ; Daniel Pape ; Caterina Petrone ; Pascal Perrier (eds.).
pages cm. - (Speech production and perception; Volume 3)
ISBN 978-3-631-66506-0 (Print) - ISBN 978-3-653-05777-5 (E-Book)
1. Speech–Psychological aspects. 2. Speech acts (Linguistics) 3. Difference (Psychology) 4. Speech perception. I. Fuchs, Susanne, 1969- editor. II. Pape, Daniel, 1975- editor. III. Petrone, Caterina, 1979- editor. IV. Perrier, Pascal.
P37.5.S68I54 2015
401'.9–dc23
2015033447

ISSN 2191-8651
ISBN 978-3-631-66506-0 (Print)
E-ISBN 978-3-653-05777-5 (E-Book)
DOI 10.3726/978-3-653-05777-5

Peter Lang Edition is an Imprint of Peter Lang GmbH.

Peter Lang – Frankfurt am Main · Bern · Bruxelles · New York · Oxford · Warszawa · Wien

This publication has been peer reviewed.

www.peterlang.com

Contents

Preface

In the night of January 1st, 2015, mankind approached a size of 7.284.283.000 human beings (see http://www.dsw.org/home.html). In this context, it seems an illusion to study individual behaviour in speech production and perception, even within a certain language. However, inter-individual variation in speech is a topic of increasing interest in linguistics, psychology, and it is the topic of our book.

Why?

Theoretical approaches have undergone a paradigm shift, moving from abstractionist to exemplar, and hybrid models. Abstractionist models treat speaker variation independently of abstract linguistic entities and consider it as noise in the data, which could be eliminated. A different view is taken by exemplar approaches assuming no separation of linguistic categories from other contextual information, e.g., indexical information about the speaker and his/her voice. All these may potentially be stored in memory. Both approaches can be seen as two extremes, but various ideas may be combined (hybrid models). In this sense we would not doubt that abstract representations of linguistic categories exist, but we would also acknowledge the richness and multidimensionality of speech signals which can facilitate speech perception.

When we talk about individual behaviour in this book, we are specifically interested in the details of the speech signals that can reveal us further insights into multiple factors affecting speech production, processing, and comprehension. So far, we are not interested in every little detail of a single speaker or listener, but rather in consistent details of speech production and perception. The crux in such an approach is to find out which of these details reveal important information about the biological, linguistic, cognitive, and social underpinnings of language in context.

The authors of this book were successful in finding several consistencies and discuss them in light of the mechanisms involved in the fascinating ability to produce and perceive speech. In particular,

Rachel Smith starts her chapter with an overview of how inter-speaker variability has been treated by different perception theories. The focus is particularly laid on abstractionist, exemplar, and hybrid approaches. These vary in how much they take into account inter-speaker variability as an information source and store this information in memory. The author continues with a comprehensive review of studies investigating fine phonetic detail which can reveal insights concerning numerous variables of a given speaker and commonalities across speaker groups.

Frank Eisner reviews some recent findings on how listeners can adapt to speaker variation and which role this variation plays for learning perceptual categories in adults. Eisner provides evidence that exposure to multiple speakers could help learning abstract representations on a lexical level. Sub-lexical processing of speaker idiosyncratic properties additionally has an impact on speech perception as shown by neurobiological and computational models. In particular, previously learned idiosyncratic properties influence perceptual expectations.

Marieke van Heugten, Christina Bergmann, and **Alejandrina Cristia** provide complementary evidence about perceptual learning with a particular focus on spoken language acquisition. Specifically, they review the literature on how young children and toddlers cope with speaker differences, regional accents, and language variation when acquiring their mother tongue. Although processing unfamiliar voices and accents is more complex than processing familiar ones, small children are extremely flexible in coping with speaker variation, and they even take advantage of it to learn their language. Indeed, infants use variability in speakers' voices to access the underlying structure. Differences in the way individuals speak can thus serve as a frame of reference to help infants accommodate variation.

Benjamin Swets studies the cognitive architecture of language. He summarizes his work on individual differences in the scope of advance planning. His results show consistently that individual differences can be systematic and, in his particular topic, reveal insights into the relation between individual working memory capacities and the scope of advance speech planning. Furthermore, he suggests that the size of working memory capacities could play a general role in packing information together for production and comprehension purposes.

Francesco Cangemi, Martina Krüger, and **Martine Grice** explicitly study the nature of the link between speaker- and listener-specific behaviour in the production and perception of prosodic categories. Their particularly novel finding is that speakers vary contextually, i. e. a given speaker can be more intelligible than other speakers for a particular listener, although she/he may be less intelligible than average for another specific listener. These findings suggest that speech comprehension of prosodic categories is shaped by the specificities of particular dyads.

Iris Chuoying Ouyang and **Elsi Kaiser,** too, dedicate their chapter to prosody. They investigate the prosodic realization of information-structural factors (new-information and corrective focus), crossed with information-theoretic factors (word frequency and contextual probability), in terms of both inter- and intra-speaker variability. The results show that these two types of factors interact in determining several aspects of the fundamental frequency contours. Moreover, speakers exhibit individual variability regarding the magnitude of prosodic cues, but the direction of prosodic distinctions between information categories is consistent across speakers.

Melanie Weirich presents her work on organic sources for inter-speaker variability in articulation with an emphasis on palatal shape, vocal tract dimensions, and tongue biomechanics. The speaker groups that are taken into account are monozygotic versus dizygotic twins who grew up together, and male versus female adults. Based on the analyses of selected phonemes and phonemic contrasts, it is shown that individual differences in organic structures can at least partially explain some idiosyncratic aspects of articulation, and the often observed speaker variation is far more than only random noise.

Pascal Perrier and **Ralf Winkler** tackle inter-speaker variation from the perspective of the biomechanical properties of the orofacial system. For this purpose they used biomechanical models, since there is no direct way to observe the consequences of the control by the Central Nervous System and those of the biomechanics of the motor system independently. In the first study, the authors show that inter-speaker differences in the main fibre direction of the Styloglossus muscle can shape the articulatory and acoustic variability in a high vowel. In the second study, the authors show that different implementations of the Orbicularis Oris muscle have an impact on the degree of lip aperture in speech production.

Jean-François Bonastre, Juliette Kahn, Solange Rossato, and **Moez Ajili** complete the book with their chapter on an applied topic – forensic speaker recognition. They particularly warn about deriving conclusions about the detection of a speaker, similarly to a fingerprint or a DNA analysis. The acoustic signal of a speaker can't be interpreted as physical biometrics. It is a complex signal including information about the human being as a bio-psychosocial unit in interaction with others. The authors summarize the main weaknesses of the methodology that make forensic phonetics in court a controversial topic, even if automatic speech recognition has substantially improved its algorithms over the last decades.

This book was inspired by the ideas from the project "SPEECHart- Speaker-specific articulation as adaptation to individual vocal tract shapes" (sponsored by the German Research Council) and the fourth summer school on „Speech production and perception: Speaker-specific behaviour", which was held from the September 30th to October 4th, 2013, in Aix-en-Provence. The summer school was jointly organized by the Laboratoire Parole et Langage in Aix-en-Provence, the Centre for General Linguistics in Berlin, and the GIPSA-lab in Grenoble. It could take place thanks to the financial support by the Ministry for Education and Research (BMBF) and the PILIOS project which was sponsored by the French-German University in Saarbrücken.

Rachel Smith
University of Glasgow

Perception of Speaker-Specific Phonetic Detail

Abstract: The individual speaker is one source among many of systematic variation in the speech signal. As such, speaker idiosyncrasies have attracted growing interest among researchers of speech perception, especially since the 1990s, when theories began to treat variation as information rather than noise. It is now a common assumption that people remember and respond to speaker-specific phonetic behaviour. But what aspects of speaker-specific behaviour are learned about and used to guide perception? Do listeners make full use of the richness of speaker-specific information available in the signal, and how can listeners' use of such information be modelled? In this chapter I review evidence that processing of the linguistic message is affected by inter-speaker variation in a number of aspects of *phonetic detail*. Phonetic detail is defined here as patterns of phonetic information that are systematically distributed in the signal and perform particular linguistic or conversational functions, but whose perceptual contribution extends beyond signalling basic phonological contrasts (such as differences between phonemes or between categories of pitch accent). Following Polysp, the *Poly*systemic *S*peech *P*erception model of Hawkins and colleagues (Hawkins and Smith, 2001; Hawkins, 2003, 2010), I argue that people can learn about speaker-specific realisations of any type of linguistic structure, from sub-phonemic features up to larger prosodic structures and, potentially, conversational units such as speaking turns. Speaker-specific attributes may even, on a more associative basis, enable direct access to aspects of meaning. I discuss circumstances liable to promote or disfavour the storage of speaker-specific phonetic detail, considering issues such as the frequency and salience of particular speaker-specific patterns in the input, and listener biases in attribution of variation to possible causes.

1. The changing role of the speaker in speech perception theories

Individual speakers are a source of considerable variability in the realisation of linguistic categories. This much has been clear since the early days of acoustic phonetics: for example, Peterson and Barney (1952) measured formant frequencies of American English vowels spoken by adult male, female and child speakers, and demonstrated not only extensive within-category variation, but also between-category overlap, when vowel tokens were plotted in F1-F2 space. Very many speech production studies show that, while speakers behave consistently with one another in many ways, there is also a significant degree of variability among them. For example, Johnson et al. (1993) found variation in the degree to which speakers of American English recruited the jaw to produce low vowels; Borden and Gay (1979) observed some speakers to produce /s/ with the tongue-tip up and others with it down (for a few more examples among many, see Dilley et al., 1996; Fougeron and Keating, 1997; van den Heuvel et al., 1996).

The implications of this inter-speaker variability for perception have been interpreted in shifting ways over the years. In the 1970s and 1980s, the dominant assumption was that speaker variability had to be stripped away, or *normalised,* before sounds and words could be recognised. Halle (1985: 101) writes: "when we learn a new word we practically never remember most of the salient acoustic properties that must have been present in the signal that struck our ears; for example, we do not remember the voice quality of the person who taught us the word or the rate at which the word was pronounced." Views such as Halle's are often referred to as *abstractionism:* i.e. the assumption that the brain must store abstract linguistic units, in order to account for the compositionality of language (e.g. McClelland and Elman, 1986; Norris et al., 2000; Pisoni and Luce, 1987). According to abstractionist views, the perceptual details of individual utterances do not ordinarily form part of linguistic representation. (Nonetheless the perceptual details of spoken utterances can be remembered and accessed for some purposes, such as autobiographical memory.) With isolated exceptions (Klatt, 1979 and to a lesser extent Wickelgren, 1969), the idea that words are stored in the form of discrete symbolic units dominated psycholinguistics and speech perception research until the 1990s. Accordingly, researchers sought to develop the best algorithms to normalise the speech signal across speakers, and/or to identify properties of sounds that remained invariant across speakers (e.g. Stevens, 1989).

From the 1990s, this view encountered a radical challenge from exemplar (also known as non-analytic or episodic) approaches to speech perception. According to these approaches (e.g. Goldinger, 1996, 1998), individual exemplars or instances of speech are retained in memory. When a new speech signal is encountered, it is matched simultaneously against all stored exemplar traces in memory, and each stored exemplar is activated in proportion to the goodness of match. The aggregate of these activations produces a response. There is no need for storage of abstract forms; linguistic categories are simply the distributions of items that a listener encounters, encoded in terms of values of parameters in a multidimensional phonetic space. Accordingly, information about the speaker need not be stripped away: it is assumed to be retained in memory, and to play a role in perception. Early work within the exemplar framework (e.g. Goldinger et al., 1991; Palmeri et al., 1993; Nygaard et al., 1994) showed that perception can be facilitated when conditions allow information about the speaking voice to be encoded and accessed (and, conversely, can be disrupted under less optimal conditions). This work emphasised global speaker characteristics like f0, vocal effort and rate (e.g. Bradlow et al., 1999; Schacter and Church, 1992; Church and Schacter, 1994; Nygaard et al., 1995).

Subsequently the pendulum swung back to a somewhat more categorical view that mixes elements of the abstractionist and exemplar approaches. This *hybrid* approach was motivated particularly by the need to explain how learning about one word may transfer to other words containing the same sound. For example, if listeners learn that a particular spectral profile is appropriate for a given speaker's /s/ in the word *mice,* they will, assuming other conditions stay sufficiently constant, expect a similar spectral profile for that speaker's /s/ in *house, dice, miss,* etc. (McQueen et al., 2006). Such patterns of generalisation across words may be difficult to explain in a purely exemplar framework, unless a degree of abstraction is assumed. Thus, Cutler et al. (2010) propose that speech is represented prelexically in terms of abstract phoneme categories, which are updated where relevant with specific information about how each phoneme is pronounced by individual speakers. Evidence supporting this position has come primarily from experiments focusing on idiosyncratic pronunciations of individual segments. A case in point is the line of research pioneered by Norris, McQueen and Cutler (2003) in which realisation of a fricative was manipulated to be ambiguous between [f] and [s]: after being exposed to the ambiguous fricative in words containing either [f] or [s] listeners shifted their perceptual category boundary between [f] and [s] to accommodate the new variant. Further research along similar lines has shown similar patterns of learning for idiosyncratic pronunciations of stops (Kraljic and Samuel, 2006) and vowels (Maye et al., 2008; Dahan et al., 2008). Based on experimental results, some researchers have proposed that the prelexical representations that undergo retuning may be allophonic rather than phonemic (e.g. Mitterer et al., 2013; Reinisch et al., 2014). However, these proposals contain little detail on questions such as how many and how subtle allophonic variants would be separately represented. Thus the idea that adaptation focuses on phonemic categories remains the most fully-developed hybrid approach.

Recently, a new class of speech perception models has emerged that deal with probabilistic processing in terms of a set of statistical concepts known as Bayesian inference (Scharenborg et al., 2005; Norris and McQueen, 2008; Clayards et al., 2008; Feldman et al., 2009). Bayes' theorem gives formal expression to the idea that under conditions of uncertainty, probabilistic inferences are made based on knowledge or expectation ('prior probability distributions') in combination with current evidence. While most Bayesian models of speech perception do not deal explicitly with speaker-related variation, Kleinschmidt and Jaeger (2015) propose a speaker-specific belief updating model, which involves inferences at multiple levels: inferences about which linguistic categories are being produced, inferences about who is speaking, and inferences about the mappings between acoustic cues and linguistic categories that the speaker is using. In Kleinschmidt and Jaeger's words (2015: 151-2), "good speech perception depends on using an appropriate generative model for the current talker, register, dialect, and so forth. The listener never has access to the true generative model, but rather only their uncertain beliefs about that generative model. Thus, adaptation can be thought of as an update in the listener's talker- or situation-specific beliefs about the linguistic generative model." The notion of a linguistic generative model is very broad and carries no commitment to any specific linguistic unit or units as the object of belief updating. However, the modelling carried out so far within this framework focuses on distributions of individual acoustic cues to phonemic contrasts, e.g. VOT as a cue to voicing or spectral centre of gravity as a cue to fricative place of articulation.

In summary, any theory of speech perception must account in some way for inter-speaker variability. Current views favour some degree of retention of speaker-specific information in memory, rather than assuming all such information is stripped away during perception. In terms of the phonetic nature of speaker-specific information that is retained, most work has focused on global prosodic attributes of a speaker, on idiosyncratic realisation of phonemes, or on speaker-specific distributions of individual cues to phonemic contrasts (see e.g. Samuel and Kraljic, 2009, for an overview). These choices may reflect either a theoretical commitment (e.g. Cutler et al., 2010), or simply be convenient for model-building. Either way, they present a rather restrictive picture of what speaker-specific behaviour can entail. The main purpose of this chapter is to argue, from phonetic and perceptual evidence, that a broader view of speaker-specific phonetics should be taken. To adopt the terms of Kleinschmidt and Jaeger (2015), this amounts to arguing that what is needed is a richer specification of the linguistic generative model about which listeners have speaker-specific beliefs.

2. Speaker-specific phonetic detail (SSPD)

Many dimensions of speaker-specific behaviour relate to linguistic structure and linguistic categories, but in ways that cannot be captured if speech is considered solely in terms of an inventory of phonemes and major intonational categories. Rather, there are dimensions of speaker-specific behaviour that involve *phonetic detail.* As defined by (among others) Local (2003) and Hawkins (Hawkins and Smith, 2001; Hawkins, 2003, 2010), phonetic detail refers to phonetic information that affects people's responses but "is not considered a major, usually local, perceptual cue for phonemic contrasts in the citation forms of lexical items" (Hawkins and Local, 2007: 181). This type of information is "systematically distributed [according to linguistic/communicative function] but not systematically treated in conventional approaches" (*ibid.*). Thus, phonetic detail refers not to information that mainly distinguishes phonemes (such as /pa/ vs. /ba/), but to cues that distinguish other aspects of linguistic structure, such as prosodic structure (compare the unstressed /p/ in *potato* with the stressed /p/ in *important*); syllabic and morphological structure (/p/ is more heavily aspirated in the morphologically-complex word *displease* than in the mono-morphemic word *displays*; Smith et al., 2012); or pragmatic function (for Standard Southern British English, both [pʰ] and [p'] are possible allophones of /p/ in *it's a tap*, but the ejective sounds more emphatic, definite, and final than the aspirated stop.

The range of aspects of linguistic structure that condition systematic variation in phonetic detail is extensive. Crucially for the present purposes, there is evidence of speaker-specific variation in many of them, henceforth termed *speaker-specific phonetic detail* (SSPD).

For example, speakers vary in the extent to which they coarticulate, and in the precise coarticulatory strategies that they use. Reviewing research in this area, Kühnert and Nolan (1999) comment that it is relatively scarce, and that "the high variability found in the data makes it difficult to distinguish between effects which should be considered as being idiosyncratic and effects which simply reflect the allowed range of variation for the phenomenon". Nonetheless, they identify several experiments showing individual coarticulatory differences: among British English speakers in coarticulation of /r/ and /l/ with a following vowel (Nolan, 1983, 1985), and among both Swedish (Lubker and Gay, 1982) and English speakers (Perkell and Matthies, 1992) in the timing of movements for anticipatory lip rounding. Some of this variation may be due to an individual's genetic (anatomical and physiological) inheritance, as suggested by Weirich et al.'s (2013) finding that tongue looping trajectories are more similar in monozygotic twins than in dizygotic twins or unrelated speakers (though see Nolan and Oh, 1996 for a demonstration of articulatory variability within identical twin pairs).

Speakers also vary in their "prosodic signatures", i.e. the detailed phonetic means they use to index prosodic prominence and prosodic boundaries. With respect to prominence, individual speakers mark prominent as opposed to non-prominent words using different subsets of prosodic properties, such as lengthening, pausing, increased intensity, increased f0, location of an f0 peak, and formant frequencies (Dahan and Bernard, 1996; Mo, 2010). With respect to prosodic boundaries, speakers vary subtly in the way they mark boundaries between syllables and words (Lehiste, 1960; Quené, 1992; Smith and Hawkins, 2012). For example, Smith and Hawkins (2012) recorded speakers of Standard Southern British English producing phonemically-identical sentence pairs, such as "*So he diced them*" vs. "*So he'd iced them*", "*They also offer Mick stability*" vs. "*They also offer mixed ability*", and found variation in patterns of allophonic detail at word boundaries: different speakers used duration to differing extents to mark the contrast between word-initial and word-final allophones, and some speakers lenited word-final sounds more than others. Similar variation occurs in the way speakers distinguish other types of prosodic domain, as shown by Redi and Shattuck-Hufnagel (2001) with respect to glottalisation, and by Fougeron and Keating (1997) for articulatory lengthening and strengthening. Across these studies, not only do different speakers preferentially use different properties to signal a distinction, but some speakers clearly distinguish all levels of the prosodic hierarchy, while others tend to "flatten" it, i. e. they fail to exploit the possible range of prosodic levels (Fougeron and Keating, 1997; Mahrt et al., 2012).

Furthermore, as outlined by Abercrombie (1967), Laver (1980) and Mackenzie Beck (2005) among others, speakers differ in their long-term settings of the larynx and the supralaryngeal articulators. These articulatory settings impart characteristic qualities that systematically colour vocal output, such as breathiness, creakiness, dentalisation, labialisation, denasalisation, and so on. In Abercrombie's description (1967: 91), such settings result in "a quasi-permanent quality running through all the sound that issues from [a person's] mouth". Importantly, however, the auditory consequences of such long-term settings depend in complex ways on the segments of the message, and also on the prosody (Mackenzie Beck, 2005). Thus if a speaker has a labialised voice quality, this will be audible on many of his/her segments, but not equally on all: segments normally produced with spread lips (e. g. /s/, /θ/, /i/) will be particularly susceptible, while segments that are ordinarily labialised may sound more extremely so (e. g. /ʃ/, /r/, /ʤ/, /ʧ/). Likewise a creaky voice quality may be especially audible at points in an utterance where creak is not normally found (e. g. phrase-medially in sonorant stretches of speech), as well as being heard as more extreme creakiness in places where creak is usual (e. g. phrase-finally, before word-final voiceless stops, between abutting vowels). By considering articulatory settings, we see that the way a speaker pronounces one of their phonemes is rarely completely independent of the way they pronounce others, yet a setting does not alter all phonemes in the same way, and prosody plays a role too.

Speakers also vary in longer-domain characteristics such as their speech rate, articulation clarity, and patterns of speech reduction (Hanique et al., 2015). Some of these longer-domain characteristics interact with the realisation of particular segments or features: Theodore et al. (2009) found that speakers vary in the extent to which changes in speech rate alter their characteristic VOT patterns. Looking beyond the prosodic hierarchy as usually defined, there are systematic patterns of phonetic detail that occur over speaking turns and other interactionally-relevant chunks of talk (see e.g. Ogden, 2012). It seems plausible that individual speakers might implement these in idiosyncratic ways, although research has not addressed this issue to date.

In summary, a speaker's phonetic individuality amounts to much more than a collection of phoneme realisations and some average prosodic properties. Speakers demonstrably vary in a number of aspects of phonetic detail, including their long-term articulatory settings, their coarticulatory behaviour, and the way they implement linguistic distinctions relating to prosodic structure. If we are accustomed to thinking about speech primarily in terms of the phonemic contrasts that distinguish individual words (e.g. *bin* vs *pin*), these types of SSPD may appear trivial, unsystematic, and of limited relevance to segment and word identification. However, when we think about recognition of words in their broader context—that is, in meaningful utterances heard in the flow of ordinary interaction—these aspects of sound structure take on a much greater importance, because they contribute some of the "glue" that holds chunks of speech together and makes them sound coherent. They help to encode phonological *structure* as well as phonological *system*; they represent "prosodies" as defined in Firthian prosodic analysis (see e.g. Ogden, 2012), or what in other phonological frameworks might be called prosody-segment interactions. If we broaden the definition of the listener's task to include grasping the semantic, grammatical, information-structural and interpersonal relations within an utterance and a conversation, we can see that the above types of phonetic detail could well play an important role in understanding the message. Therefore, there is a clear potential advantage for listeners in learning to interpret patterns of SSPD produced by individual familiar speakers. The next section discusses whether listeners do in fact learn about and use these types of SSPD.

3. Evidence for use of SSPD in speech perception

If listeners know about speaker-specific phonetic detail as defined above — as opposed to simply about how a speaker realises their phonemes, or about their average vocal pitch, etc. — and if they use this knowledge in perception, two consequences can be expected. First, exposure to a person's speech should lead to changes in task performance that are *general* to properties that are common across groups of sounds a speaker produces. For example, if listeners are exposed to Fred's voiceless alveolar plosives, which are dentalised and have extremely long VOT, they may form expectations that Fred will produce other voiceless plosives with long VOT, and/or that he will dentalise other alveolar sounds. Thus responses to Fred's long-VOT /k/ and /p/, and to his dentalised /d/ and /n/, should be primed (facilitated) by the prior exposure to his tokens of /t/. Second, exposure to a person's speech should lead to changes in task performance that are *specific* to sounds that occur in particular structures. If Jill lenites word-final /d/ to an unusual extent, realising it as an approximant in unstressed function words like *he'd* and *she'd,* listeners might expect similarly lenited variants in her unstressed *we'd* and *I'd,* and possibly also in stressed tokens of these words and in stressed content words; but they would not necessarily expect to hear them in Jill's pronunciation of word-initial /d/. If, on the other hand, listeners only adjust phoneme categories when accommodating perceptually to a speaker, simpler patterns of responding would be expected. Exposure to Fred's /t/s should only affect subsequent responses to /t/, and not to /p/, /k/, etc.; while exposure to Jill's word-final /d/ in *he'd* should affect responses to /d/ in all other contexts.

These questions about generalisation and specificity in perceptual learning about individual speakers have been addressed experimentally by a small number of studies. Some of these explicitly manipulate both linguistic structure and speaker identity: that is, they use multiple speakers, and test whether listeners learn to associate a structure-specific pattern with an individual speaker, and not with other speakers who do not produce the pattern. While such experiments represent the "gold standard", they are quite hard to conduct, as the requirement to test listeners with multiple voices *and* multiple linguistic structures leads to very long experiments. Therefore, a shortcut is sometimes taken: Experiments test whether an unusual pattern can be learned from a single speaker's voice. If it can, it is inferred that in adapting to this unusual type of speech, listeners have learned a potentially speaker-specific property (e.g. Barden and Hawkins, 2013; Poellmann et al., 2014). Clearly, only the former type of experiment *directly* demonstrates perceptual use of SSPD. Nonetheless, the inference drawn from the latter type is probably valid. Dahan et al. (2008) used one voice to assess perceptual learning of a contextually-conditioned allophone, and inferred from their results that the learning might be speaker-specific; Trude and Brown-Schmidt (2012) conducted a similar experiment with multiple voices, and confirmed direct evidence of speaker-specific learning.

There is evidence for the first claim, i.e. for perceptual use of SSPD relating to *general* properties that are common across groups of sounds a speaker produces. This evidence relates to the feature [±voice]. Individual speakers differ in their characteristic VOT in voiceless stops (Allen, Miller, and DeSteno, 2003). Listeners can learn to associate a speaker with a characteristic pattern of VOT (Allen and Miller, 2004; Theodore and Miller, 2010; though under some circumstances learning about the realisation of [±voice] may generalise to other speakers, Kraljic and Samuel, 2007). Several studies using a range of learning paradigms have shown that speaker-specific learning about VOT generalises not only among words beginning with the same phoneme (e.g. /t/), but also, partially or fully, across place of articulation, i.e. to other voiceless stop phonemes (Kraljic and Samuel, 2006; Theodore and Miller, 2010; Nielsen, 2011).

There is also evidence for the second claim, i.e. that exposure to a person's speech can lead to changes in task performance that are *specific* to sounds that occur in particular contexts or structures. Several studies address whether learning about how a speaker pronounces a sound in one position in the syllable or the word generalises to other positions. Smith and Hawkins (2012) tested the perceptual relevance of the individual differences in phonetic detail at word boundaries discussed in section 2 above. Tests of intelligibility in noise before and after exposure to a voice showed that familiarity with an individual speaker's patterns helped listeners to segment and identify words in noise. The learning was speaker-specific, and the perceptual benefit was small, but robust. Some other work on transfer of speaker-specific learning across positions in syllable or word supports Smith and Hawkins' findings: Dahan and Mead (2010) found, for a range of phonemes, that learning to understand noise-vocoded speech was specific to position in syllable. However, Jesse and McQueen (2011) found that learning of an unusual pronunciation of /s/ was not specific to position in syllable. The divergent results may be due to the different phonemes under investigation, and/or to other aspects of the experiments. For example, if the critical acoustic information for the perception of /s/ is contained mainly within the fricative itself, rather than distributed across more than one segment, this may encourage generalisation across positions (Reinisch et al., 2014). Relatedly, Jesse and McQueen (2011) spliced the identical fricative across positions in syllable, whereas the syllable-initial and -final fricatives in Smith and Hawkins' study exhibited natural variation in duration and spectral composition.

Other research shows perceptual learning of speaker-specific phonetic detail that relates to specific allophones rather than specific phonemes. Dahan et al. (2008) exposed listeners to a dialect in which /æ/ is raised to [eɪ] or [ɛ] before voiced velar stops (e.g. in *bag*) but not voiceless ones (e.g. *back*). They hypothesized that if listeners learned this pattern, they would obtain an advantage in recognising the words: they would be able to use the information in the vowel to resolve the lexical competition between *bag* and *back* earlier in the time course of the word. Listeners' eye-tracking performance supported this hypothesis: listeners who had been exposed to the raised vowel identified *bag,* as opposed to its competitor *back*, earlier and more accurately than listeners who had been exposed to the standard variant of the vowel. Trude and Brown-Schmidt (2012) replicated Dahan et al's finding, varying the voice heard in the test phase and thereby demonstrating that the learning was genuinely speaker-specific. A different type of allophonic variation was shown to be perceptually important by Mitterer, Scharenborg and McQueen (2013). They generated an ambiguous segment by averaging approximant /r/ and dark /l/. Learning about this ambiguous segment altered performance on an approximant-/r/-to-dark-/l/ continuum, but not on continua where the endpoints were trill /r/ and light /l/.

In another test of the structure-specificity of perceptual learning, Barden and Hawkins (2013) investigated perceptual learning of phonetic patterns related to morphological structure. Grammatical morphemes can be pronounced differently from the identical phoneme strings when these do not function as morphemes. For example, the phoneme sequence /mɪst/, when spoken in a prefixed word like *mistimes,* has a longer and more peripheral /ɪ/, shorter /s/, and a /t/ with longer VOT, than when spoken in a word that does not have a true prefix, such as *mistakes* or *mystique* (Smith, Baker and Hawkins, 2012). The *re-* of *repaint* (which decomposes morphologically into *re + paint*) likewise has a more peripheral vowel than the *re-* of *report* (which does not decompose into *re + port*). Barden and Hawkins asked whether, if exposed to an idiosyncratic pronunciation of a prefix, listeners would learn to expect this pronunciation in prefixed but *not* non-prefixed words. They trained two groups of listeners with stories containing prefix *re-*, either realised unusually as /rɪ/ (Accent group), or realised normally as /ri/ (Control). Listeners then performed an intelligibility-in-noise test containing keywords with prefix *re-* (e.g. *republication*) and non-prefix *re-* (e.g. *renal infection*) pronounced as /rɪ/. Listeners in the Accent group, who had been exposed to the /rɪ/ prefix, identified the unusually-pronounced keywords significantly more accurately than listeners in the Control group. The benefit was present for both prefixed and non-prefixed test words, but was significantly greater for prefixed words, suggesting the listeners associated the unusual pronunciation more strongly with the specific linguistic structure in which it had been encountered, though the learning did partially generalise to other structures.

Along similar lines, Poellmann et al. (2014) demonstrated that listeners could adapt to particular realisations of a prefix that are characteristic of fast casual speech. Listeners who were exposed to words beginning with the Dutch prefix *ver-*, realised as [fː], showed improved identification of new *ver-* words realised with [fː], compared to unexposed listeners. Listeners in this study may have been learning about prefix pronunciation, or speech style, or both. The data do not allow these possibilities to be distinguished, but regardless, they underscore that perceptual learning cannot solely concern phonemic categories.

At the other end of the spectrum, there is also evidence from a different line of research, that what listeners learn about a voice is restricted neither to its gross prosodic properties nor its segmental fine structure. Several experiments have investigated perceptual learning by applying different types of degradation to the speech signal. Remez et al. (1997), Remez et al. (2002) and Sheffert et al. (2003) used sine-wave speech, which lacks natural vocal quality and segmental-phonetic fine structure, but preserves enough of the time-varying spectro-temporal structure of speech to support word recognition. Adult listeners are surprisingly good at identifying personally familiar talkers from sine-wave replicas of their utterances (Remez et al., 1997). Moreover, adults can generalise knowledge of speaker-specific attributes that has been learned from sine-wave replicas to both novel sine-wave samples and natural speech (Sheffert et al., 2003). Interestingly, pre-school children can also discriminate familiar cartoon voices from spectrally-degraded (in this case noise-vocoded) speech (Van Heugten et al., 2014). The acoustic basis for speaker identification from degraded speech samples is not yet clear, but it presumably must rely partly on global qualitative speaker characteristics such as formant spacing, which are preserved in sine-wave and (given sufficient spectral resolution) noise-vocoded speech. Local phonetic properties (such as segmental durations) probably also play a role, but their specific importance has not been tested.

In summary, listeners can learn many aspects of SSPD. Learning sometimes transfers across phoneme categories, as in the case of VOT in voiceless stops. Learning does not necessarily transfer to all members of a phoneme category: it may be specific to certain positions in word, or to certain morpho-lexical structures, such as prefixes. From the evidence so far, it is reasonable to assume that the patterns of transfer are not arbitrary, but principled, reflecting how general vs. how specific to particular linguistic structures the phonetic properties in question actually are.

4. Modelling the perceptual relevance of SSPD

What kind of a model can account for the data on perception of speaker-specific phonetic detail — i.e. for listeners' ability to learn patterns that are specific *both* to an individual speaker *and* to a particular (type of) linguistic context? The preceding sections have shown that models that assume abstract phoneme categories, updated with speaker-specific information (e.g. Cutler et al., 2010), cannot fully do so, because some aspects of SSPD generalise across phoneme categories, while others are restricted to only some instances of a phoneme category. At the same time, some abstraction is needed, to account for the patterns of generalisation that have been shown in perceptual learning studies, as well as to explain why exemplar effects are not consistently found across all experiments (see e.g. Hanique et al., 2013).

Smith and Hawkins (2012) discuss a range of modelling approaches that have the potential to accommodate their data on speaker-specific word segmentation. Here, I focus on Polysp (Hawkins and Smith, 2001; Hawkins,

2003, 2010), which is not a computationally implemented testable model, but indicates the lines along which such a model could develop. Polysp stands for *Polysystemic Speech Perception*; the term 'polysystemic' reflects the idea that the phonology of a language involves a range of structures, within each of which different systems of contrast may operate, as opposed to a single monolithic phoneme system (Hawkins and Smith, 2001). The model takes a hybrid episodic-abstract approach, and posits that phonetically detailed episodes are stored in memory alongside abstraction in terms of rich linguistic structures. Exhaustive parsing of the signal into abstract linguistic categories is argued not to be needed if meaning can be accessed without it. This may be the case when listeners hear familiar chunks of highly reduced speech: for example, [ə̰̃ə̃ə̰̃] is (in some circumstances) an acceptable, if highly casual, realisation of the phrase *I don't know,* and probably does not need to be mapped on to the three individual words in order for the listener to understand that the speaker lacks some knowledge or information. Access to meaning without parsing into abstract categories may also occur in situations where identifying a particular voice is sufficient to constrain the interpretation of a linguistic structure or meaning, as demonstrated in an eye-tracking study by Creel and Tumlin (2011). Nonetheless, in general, Polysp proposes that phonological knowledge is represented in terms of rich, hierarchical structures (incorporating prosodic and grammatical information) and this representation improves the process of pattern-matching between signal and memories. These structures are abstract (like phonemes) but are richer than a phoneme string, and as such allow for more complex phonetic detail to be represented. Speaker-specific information can potentially be associated with any unit(s) at any level(s) of the representation.

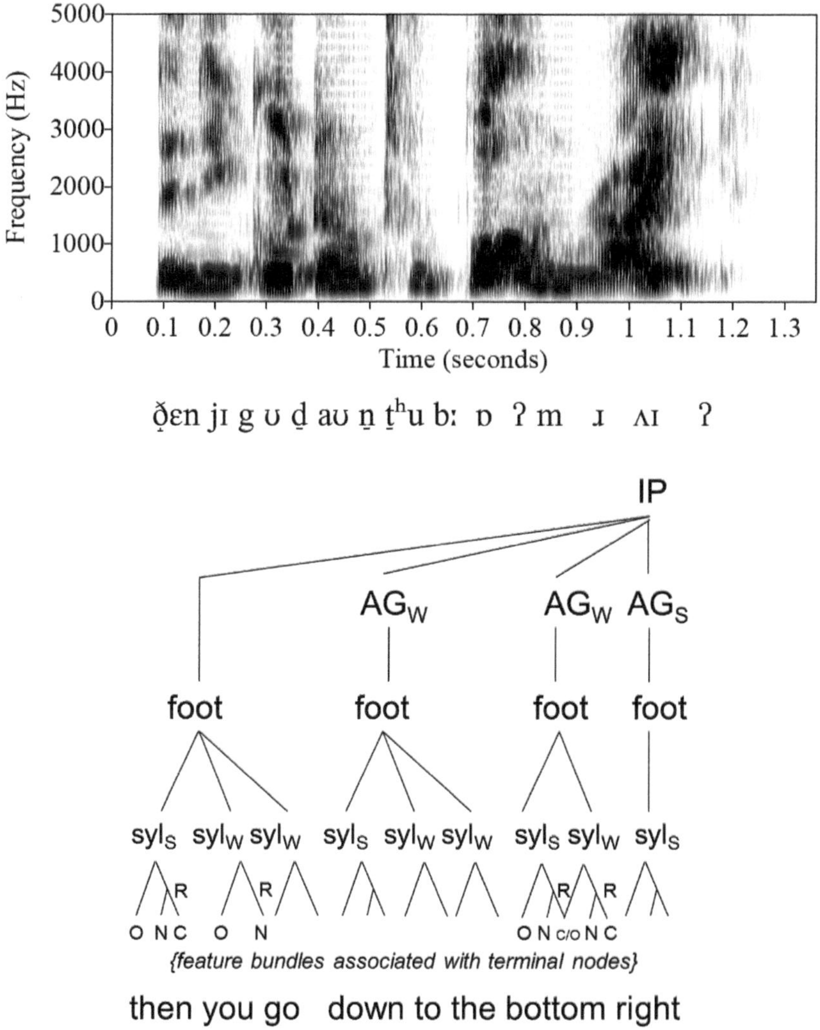

Figure 1. The utterance then you go down to the bottom right, *spoken by a young male Panjabi-English bilingual speaker from Bradford (taken from the IViE corpus, www.phon.ox.ac.uk/IViE/). Top panel: Wideband spectrogram and phonetic transcription of the utterance. Bottom panel: Representation of the utterance as a prosodic tree. IP = Intonational Phrase; AG = Accent Group; S = strong, W = weak; O = onset, R = rime, N = nucleus, C = coda. Each terminal node in the tree could further be associated with a bundle of distinctive features, not represented here.*

Figure 1 represents an utterance spoken by a male teenage bilingual Panjabi-English speaker from Bradford (UK), performing a map task (taken from the IViE corpus, www.phon.ox.ac.uk/IViE/). The top panel shows a spectrogram and associated phonetic transcription, while the bottom panel shows a prosodic tree corresponding to the utterance. Different theoretical approaches would differ as to the details of the prosodic tree (e.g. Selkirk, 1986; Nespor and Vogel, 1986), but this does not matter for our purposes: the main point of the tree is to show that syllabic and prosodic structures are core to this representation of the utterance, while the phoneme string is not.

The prosodic tree gives a window on the opportunities afforded by the example utterance to learn about speaker-specific phonetic detail, that is rather different from the picture presented by a phoneme string. For example, the phonemic transcription, /ðɛn jə go daʊn tə ðə bɒtəm raɪt/, indicates that the utterance contains three instances of the phoneme /t/. However, the narrow transcription and the spectrogram indicate that the speaker realises these in quite different ways, with an aspirated /t/ in *to* and glottal stops in *bottom* and *right*. This much may seem fairly banal—/t/ is well known to have considerable allophonic variation in English, with glottal stop prominent among the variants. What the tree also shows but the phoneme string does not, however, is the structural constraints on this speaker's use of glottal stop. He uses it word-medially (at the juncture between a stressed and a following unstressed syllable in *bottom*), and word-finally (in *right*), but not word-initially (*to*). A different speaker might also use glottal stop for /t/ word-initially but foot-medially (in *down to the*). A different person again might only use it word-finally.

The spectrogram also indicates that the speaker produces word-initial voiced stops quite consistently, regardless of their place of articulation: the initial stops in *go*, *down, bottom* are all voiced throughout their closures. Moreover, as the narrow transcription indicates, he produces slightly retracted alveolar consonants in *down* and *to* (which is a typical feature for British Panjabi speech; cf Alam and Stuart-Smith, 2011 and Kirkham, 2011). He reduces weak syllables quite substantially: they are considerably lower in intensity than adjacent strong syllables, and are segmentally reduced, with a syllabic nasal in *bottom* and a very minimal trace of *the* in *down to the* (the word is realised merely as some extra duration on the /b/ of *bottom*).

In summary, even from the single utterance represented in Figure 1, it can be seen that the prosodic tree makes it possible to capture a number of systematic patterns which are not evident from a segmental transcription alone. Although the structures look complex at first sight, their value from the perspective of modelling SSPD is that they allow a great deal of information about the speaker to be represented, which has the potential to predict the speaker's future behaviour in some detail.

The foregoing discussion suggests that listeners can construct speaker-specific representations that are highly detailed and complex, comprising knowledge of speaker variation at many linguistic levels. It seems reasonable to assume that full representations of this kind could be built up only with considerable exposure to a speaker's voice. That is, for a highly familiar speaker, such as a partner, parent, or close friend, the listener's stored representations could well be elaborated with probabilistic knowledge of the speaker's typical patterns at many or all of these levels. For example, a listener highly personally familiar with the speaker in Figure 1 might have detailed knowledge of how the speaker produces syllable-initial /t/ in foot-medial weak syllables, such that the listener would be surprised if this speaker were to use a glottal stop for /t/ in *down to the shops*. But when a listener is merely casually familiar with a speaker, or is beginning to get to know them, the listener would not be familiar with all the systematics of the speaker's idiolect. The speaker-specific representations would be much less detailed, and would support less confident predictions and inferences during speech understanding. Moreover, different listeners might construct quite different speaker-specific representations, because the complexity of the structures allows flexibility in mapping of phonetic patterns to representations. From limited data such as the utterance in Figure 1, a listener might abstract a generalization about how the speaker produces the phoneme /t/, or voiceless stops in general, or voiceless stops in weak syllables, and so on. There are numerous possibilities for the attribution of phonetic variation to causes[1], and only with appropriate exposure would the listener be able to disentangle these from one another and fine-tune their model of the speaker's behaviour.

In this regard, a particularly interesting set of results was obtained by Eisner et al. (2013). They looked at word-final devoicing, i.e. the pronunciation of (for example) *overload* as *overloat,* which is a common pattern in Dutch, among other languages, but occurs to a much lesser extent in English. When native English listeners were exposed to a Dutch speaker devoicing word-final /d/, their perceptual responses showed overgeneralization across positions in syllable: that is, they became more willing to accept the speaker's devoiced tokens as instances of /d/, not only in final position, but also in initial position, as in *down* pronounced as *town.* Interestingly, however, this overgeneralization did not occur if listeners were also exposed to the Dutch speaker's actual (voiced) variants of initial /d/, nor when the stimuli were presented in a native English accent. These findings underscore the flexibility inherent in learning of speaker-specific pronunciation patterns. Learning generalised across positions in syllable when listeners had no reason not to expect a speaker to produce the same variant in all contexts (i.e. in the case of the unfamiliar Dutch accent). But learning failed to generalise when listeners were presented with direct evidence of the speaker's allophonic variation (in the case where they heard the Dutch speaker producing voiced initial /d/ and devoiced final /d/). Learning also failed to generalise when listeners had a strong expectation about the patterns of *normal* allophonic variation in the variety they were hearing, as in the case where they heard the native English speaker: listeners know that native English speakers sometimes devoice word-final stops, but rarely word-initial ones, and did not show overgeneralization in this case.

Polysp does not make detailed predictions about *how* speaker-specific representations might be built up over the course of exposure, focusing

---

1 The issue here is reminiscent of the problem of the indeterminacy of translation, as discussed by Quine (1960). If we see a rabbit, and hear a speaker of an unknown language say "gavagai," there are numerous possible meanings: e.g. *Look, a rabbit. Look, food. Let's go hunting. There will be a storm tonight. Look, a momentary rabbit-stage. Look, an undetached rabbit-part.* See Kraljic et al. (2008) for a similar point about attribution of phonetic variation to causes, memorably illustrated using a Benny Hill joke.

rather on the form such representations might eventually take. However, a Bayesian model like that of Kleinschmidt and Jaeger (2015) could generate empirically testable predictions about how representations develop through exposure, if the model were expanded to express richer linguistic structure. In addition to mere exposure, there appear to exist cognitive biases which affect how a listener builds up a representation of a speaker's behaviour. Kraljic and colleagues carried out an elegant series of experiments exploring the circumstances under which listeners are willing to interpret phonetic variation as speaker-specific. They found that an unusual pronunciation was more likely to be assumed to be speaker-specific if it could not plausibly be attributed to the phonetic context (Kraljic, Brennan and Samuel, 2008), or to an extraneous proximal cause (such as a pen in the speaker's mouth; Kraljic, Samuel and Brennan, 2008). Moreover, a pronunciation was more likely to be attributed to speaker-specific behaviour if it was first encountered early on in exposure to the speaker: listeners seemed to assume that a speaker-specific characteristic should be stable, and thus if a pattern had not been encountered early in exposure, they preferred to attribute it to some more transient cause (Kraljic, Samuel and Brennan, 2008). Again, for a fuller consideration in a Bayesian framework of circumstances under which listeners may incline to rely on existing beliefs vs. develop new speaker-specific ones, see Kleinschmidt and Jaeger (2015).

In summary, hybrid models like Polysp allow perceptual learning of SSPD to be conceptualised in terms of speaker-specific modulations of rich linguistic (phonological, prosodic, grammatical) structural representations. These representations have the potential to account for some of the more complex perceptual responses to speaker-specific phonetic detail, which are harder to capture in phoneme-based models. However, the richness of the representations does create the potential for indeterminacy in attribution of phonetic patterns to causes. Various cognitive biases may be involved in resolving such indeterminacy, and more work is needed to understand these. Sufficient exposure must surely be needed — listeners cannot learn a pattern unless they hear it, obviously — but beyond this, it may be the case that some regularities are easily learnable, while others are more resistant to perceptual learning (similar arguments are made in research on the transmission of sound change: Milroy, 2007). I speculate that listeners will be better able to learn about SSPD in chunks of speech that are rhythmically and prosodically salient, and predictable in terms of meaning, because meaning is known to guide perceptual learning (Davis et al., 2005). A more general prediction is that listeners may also vary in exactly what and how they abstract from a person's speech: that is, we might expect listener-specific perception of speaker-specific phonetic detail. A listener's ability and readiness to make and generalise speaker-specific perceptual adjustments in this way might even correlate with the degree of phonetic shift (accommodation) they produce in response to a conversation partner's speech. These speculations remain to be tested empirically.

5. Conclusions and future directions

The present review has shown that speakers vary in the way they realise many complex aspects of linguistic structure, from coarticulation through context-conditioned allophony to marking of syllable and word boundaries, and casual speech reduction strategies. These patterns of individual variation can be learned about, and can facilitate performance in various laboratory tasks. A promising approach to modelling them is using hybrid models that assume some degree of exemplar or episodic storage, combined with flexible abstraction that allows speaker-specific attributes to be associated with any level of hierarchically-organised phonetic and prosodic structure.

Where might the study of the perceptual role of speaker-specific phonetic detail head next? First, more work is needed to develop models that make concrete predictions about how representations of speaker-specific phonetic detail are built up as a function of experience. Second, a critical approach to the concept of the speaker itself will help to move the field forward. The discussion so far has implicitly assumed i) that individual speakers behave stably in their production of any given linguistic structure, and ii) that the individual speaker is the main locus of interesting variation. Both these assumptions are almost certainly incorrect. Many factors contribute to variation *within* a speaker (such as his/her temporary physical and emotional state, the physical speaking/listening environment, the task he/she is engaged in, the structural constraints of conversation, and intersubjective aspects such as his/her affiliation with an interlocutor). Moreover, speakers are not islands, but cluster according to numerous variables (including sex and gender, age, personality, regional accent, socio-economic status, occupation, participation in communities of practice, and so on). Thus an understanding of the perceptual "speaker space" must ultimately take into account both variation within a speaker, and commonality across groups of speakers who share similar personal or social characteristics.

Finally, an interesting avenue to explore is how speaker-specific phonetic detail simultaneously contributes both to listeners' understanding of the linguistic message (in a lexical/linguistic 'search space'), and also to recognition of a speaker's individual identity and/or group affiliations (in 'speaker space'). The interactions between these two domains have not been thoroughly explored (though see Mullennix and Pisoni, 1990, and Creel and Tumlin, 2011 for promising directions), and many outstanding questions remain about how the tasks of speaker identification and word identification are solved in parallel, in real time. For the future, the study of speaker-specific phonetic detail can be expected to play an important role in developing an integrated account of how listeners simultaneously perceive speakers' personal and social characteristics, *and* their verbal messages.

References

Abercrombie, D. (1967). *Elements of general phonetics*. Edinburgh: Edinburgh University Press.

Alam, F. and Stuart-Smith, J. (2011). Identity and ethnicity in /t/ in Glasgow-Pakistani high-school girls. In *Proceedings of the XVIIth International Congress of Phonetic Sciences*, pp. 216–219.

Allen, J. S., and Miller, J. L. (2004). Listener sensitivity to individual talker differences in voice-onset time. *The Journal of the Acoustical Society of America, 116*, 3171–3183.

Allen, J. S., Miller, J. L., and DeSteno, D. (2003). Individual talker differences in voice-onset-time. *The Journal of the Acoustical Society of America, 113*, 544–552.

Barden, K. and Hawkins, S. (2013). Perceptual learning of phonetic information that indicates morphological structure. *Phonetica, 70*, 323–342.

Borden, G. and Gay, T. (1979). Temporal aspects of articulatory movements for /s/-stop clusters. *Phonetica, 36*, 21–31.

Bradlow, A., Nygaard, L. C., and Pisoni, D. B. (1999). Effects of talker, rate and amplitude variation on recognition memory for spoken words. *Perception and Psychophysics, 61*, 206–219.

Church, B. A., and Schacter, D. L. (1994). Perceptual specificity of auditory priming: Implicit memory for voice intonation and fundamental frequency. *Journal of Experimental Psychology: Learning, Memory, & Cognition, 20*, 521–533.

Clayards, M., Tanenhaus, M.K., Aslin, R.N. and Jacobs, R.A. (2008). Perception of speech reflects optimal use of probabilistic speech cues. *Cognition, 108*, 804–809.

Creel, S. C., and Tumlin, M. A. (2011). On-line acoustic and semantic interpretation of talker information. *Journal of Memory and Language, 65*, 264–285.

Cutler, A., Eisner, F., McQueen, J. M., and Norris, D. (2010). How abstract phonemic categories are necessary for coping with speaker-related variation. In C. Fougeron, B. Kühnert, M. D'Imperio, and N. Vallée (Eds.), *Laboratory phonology 10: Variability, phonetic detail and phonological representation* (pp. 91–111). Berlin: de Gruyter.

Dahan, D., and Bernard, J.-M. (1996). Interspeaker variability in emphatic accent production in French. *Language and Speech, 39*, 341–374.

Dahan, D., Drucker, S. J., and Scarborough, R.A. (2008). Talker adaptation in speech perception: adjusting the signal or the representations? *Cognition, 108*, 710–718.

Dahan, D., and Mead, R. L. (2010). Context-conditioned generalization in adaptation to distorted speech. *Journal of Experimental Psychology: Human Perception and Performance, 36*, 704–728.

Davis, M. H., Johnsrude, I. S., Hervais-Adelman, A., Taylor, K., and McGettigan, C. (2005). Lexical information drives perceptual learning of distorted speech: Evidence from the comprehension of noise-vocoded sentences. *Journal of Experimental Psychology: General, 134*, 222–241.

Dilley, L., Shattuck-Hufnagel, S. and Ostendorf, M. (1996). Glottalisation of word-initial vowels as a function of prosodic structure. *Journal of Phonetics, 24*, 423–444.

Eisner, F., Melinger, A., and Weber, A. (2013). Constraints on the transfer of perceptual learning in accented speech. *Frontiers in Psychology, 4*, 148.

Feldman, N.H., Griffiths, T.L. and Morgan, J.L. (2009). The influence of categories on perception: Explaining the perceptual magnet effect as optimal statistical inference. *Psychological Review, 116*, 752-782.

Fougeron, C., and Keating, P. A. (1997). Articulatory strengthening at edges of prosodic domains. *Journal of the Acoustical Society of America, 101*, 3728–3740.

Goldinger, S.D. (1996). Words and voices: Episodic traces in spoken word identification and recognition memory. *Journal of Experimental Psychology: Learning, Memory, and Cognition*, 22, 1166–1183.

Goldinger, S.D. (1998). Echoes of echoes? An episodic theory of lexical access. *Psychological Review, 105*, 251–579.

Goldinger, S.D., Pisoni, D.B. and Logan, J.S. (1991). On the nature of talker variability effects on recall of spoken word lists. *Journal of Experimental Psychology: Learning, Memory, and Cognition, 17*, 152–162.

Halle, M. (1985). Speculation about the representation of words in memory. In V. Fromkin (Ed.), *Phonetic linguistics* (pp. 101–114). New York: Academic Press.

Hanique, I., Aalders, E. and M. Ernestus (2013). How robust are exemplar effects? *The Mental Lexicon, 8*, 269–294.

Hanique, I., Ernestus, M. and Boves, L. (2015). Choice and pronunciation of words: Individual differences within a homogenous group of speakers. *Corpus Linguistics and Linguistic Theory, 11*, 161–185.

Hawkins, S. (2003). Roles and representations of systematic fine phonetic detail in speech understanding. *Journal of Phonetics, 31*, 373–405.

Hawkins, S., and Local, J. (2007). Sound to sense: Introduction to the special session. In *Proceedings of the XVIth International Congress of Phonetic Sciences*, pp. 181–184.

Hawkins, S. (2010). Phonetic variation as communicative system: Perception of the particular and the abstract. In C. Fougeron, B. Kühnert, M. d'Imperio, and N. Vallée (Eds.), *Laboratory phonology 10: Variability, phonetic detail and phonological representation* (pp. 479–510). Berlin: Mouton de Gruyter.

Hawkins, S., and Smith, R. H. (2001). Polysp: A polysystemic, phonetically-rich approach to speech understanding. *Italian Journal of Linguistics-Rivista di Linguistica, 13*, 99–188.

Jesse, A., and McQueen, J.M. (2011). Positional effects in the lexical retuning of speech perception. *Psychonomic Bulletin and Review, 18,* 943–950.

Johnson, K., Ladefoged, P. and Lindau, M. (1993). Individual differences in vowel production. *The Journal of the Acoustical Society of America, 94*, 701–714.

Kirkham, S. (2011). The acoustics of coronal stops in British Asian English. In *Proceedings of the XVIIth International Congress of Phonetic Sciences*, pp. 1102–1105.

Klatt, D. H. (1979). Speech perception: a model of acoustic-phonetic analysis and lexical access. *Journal of Phonetics, 7*, 279–312.

Kleinschmidt, D.F. and Jaeger, T.F. (2015). Robust speech perception: Recognize the familiar, generalize to the similar, and adapt to the novel. *Psychological Review, 122,* 148–203.

Kraljic, T. and Samuel, A. G. (2006). Generalization in perceptual learning for speech. *Psychonomic Bulletin and Review, 13,* 262–268.

Kraljic, T., and Samuel, A. G. (2007). Perceptual adjustments to multiple speakers. *Journal of Memory and Language, 56*, 1–15.

Kraljic, T., Brennan, S.E., and Samuel, A.G. (2008). Accommodating variation: Dialects, idiolects, and speech processing. *Cognition, 107,* 51–81.

Kraljic, T., Samuel, A.G., and Brennan, S.E. (2008). First impressions and last resorts: How listeners adjust to speaker variability. *Psychological Science, 19,* 332–338.

Kühnert, B. and Nolan, F. (1999). The origin of coarticulation. In W.J. Hardcastle and N. Hewlett (Eds.), *Coarticulation: Theory, data and techniques* (pp. 7–30). Cambridge: Cambridge University Press.

Laver, J. (1980). *The phonetic description of voice quality.* Cambridge: Cambridge University Press.

Lehiste, I. (1960). An acoustic–phonetic study of internal open juncture. *Phonetica, 5* (Suppl.), 5–54.

Local, J. (2003). Variable domains and variable relevance: interpreting phonetic exponents. *Journal of Phonetics*, 31, 321–339.

Lubker, J., and Gay, T. (1982). Anticipatory labial coarticulation: Experimental, biological, and linguistic variables. *The Journal of the Acoustical Society of America, 71,* 437–448

Mackenzie Beck, J. (2005). Perceptual analysis of voice quality: The place of vocal profile analysis. In W.J. Hardcastle and J. Mackenzie Beck (Eds.), *A figure of speech: A Festschrift for John Laver* (pp. 285–322). Mahwah: Erlbaum.

Mahrt, T., Cole, J., Fleck, M., and Hasegawa-Johnson, M. (2012). Modeling speaker variation in cues to prominence using the Bayesian information criterion. In *Proceedings of Speech Prosody, 2012.*

Maye, J., Aslin, R. N., and Tanenhaus, M. K. (2008). The weckud wetch of the wast: Lexical adaptation to a novel accent. *Cognitive Science, 32,* 543–562.

McClelland, J.L. and Elman, J.L. (1986). The TRACE model of speech perception. *Cognitive Psychology, 18*, 1–86.

McQueen, J. M., Cutler, A., and Norris, D. (2006). Phonological abstraction in the mental lexicon. *Cognitive Science*, *30*, 1113–1126.

Milroy, L. (2007). Off the shelf or under the counter? On the social dynamics of sound changes. In *Studies in the History of the English Language III: Managing Chaos: Strategies for Identifying Change in English.* Berlin: Mouton de Gruyter.

Mitterer, H., Scharenborg, O., and McQueen, J. M. (2013). Phonological abstraction without phonemes in speech perception. *Cognition, 129,* 356–361.

Mo, Y. (2010). *Prosody production and perception with conversational speech.* Unpublished Ph.D. dissertation, University of Illinois.

Mullennix, J.W. and Pisoni, D.B. (1990). Stimulus variability and processing dependencies in speech perception. *Perception and Psychophysics, 47*, 379–390.

Nespor, M. and Vogel, I. (1986). *Prosodic phonology.* Dordrecht: Foris.

Nielsen, K. (2011). Specificity and abstractness of VOT imitation. *Journal of Phonetics, 39,* 132–142.

Nolan, F. (1983). *The phonetic bases of speaker recognition.* Cambridge: Cambridge University Press.

Nolan, F. (1985). Idiosyncrasy in coarticulatory strategies. *Cambridge papers in Phonetics and Experimental Lingustics, 4,* 1–9.

Nolan, F. and Oh, T. (1996). Identical twins, different voices. *Forensic Linguistics*, 3, 39–49.

Norris, D., McQueen, J.M. and Cutler, A. (2000). Merging information in speech recognition: feedback is never necessary. *Behavioral and Brain Sciences, 23*, 299–370.

Norris, D., McQueen, J. M., and Cutler, A. (2003). Perceptual learning in speech. *Cognitive Psychology*, *47*, 204–238.

Norris, D. and McQueen, J.M. (2008). Shortlist B: A Bayesian model of continuous speech recognition. *Psychological Review, 115*, 357–395.

Nygaard, L. C., Sommers, M. S., and Pisoni, D. B. (1994). Speech perception as a talker-contingent process. *Psychological Science*, *5*, 42–46.

Nygaard, L.C., Sommers, M.S. and Pisoni, D.B. (1995). Effects of stimulus variability on perception and representation of spoken words in memory. *Perception and Psychophysics, 57*, 989–1001.

Ogden, R. (2012). Prosodies in conversation. In O. Niebuhr (Ed.), *Prosodies – Context, function and communication* (pp. 201–218). Berlin: Mouton de Gruyter.

Palmeri, T.J., Goldinger, S.D. and Pisoni, D.B. (1993). Episodic encoding of voice attributes and recognition memory for spoken words. *Journal of Experimental Psychology: Learning, Memory, and Cognition, 19*, 309–328.

Perkell, J. S. and Matthies, M. L. (1992). Temporal measures of anticipatory labial coarticulation for the vowel /u/: Within- and cross-subject variability. *The Journal of the Acoustical Society of America*, 91, 2911–2925.

Peterson, G. E. and Barney, H. L. (1952). Control methods used in a study of the vowels. *The Journal of the Acoustical Society of America, 24*, 175–184.

Pisoni, D.B. and Luce, P.A. (1987). Acoustic-phonetic representation in word recognition. *Cognition, 25*, 21–52.

Poellmann, K., Bosker, H.R., McQueen, J.M. and Mitterer, H. (2014). Perceptual adaptation to segmental and syllabic reductions in continuous spoken Dutch. *Journal of Phonetics, 46*, 101–107.

Quené, H. (1992). Durational cues for word segmentation in Dutch. *Journal of Phonetics*, 20, 331–350.

Quine, W. V. O. (1960). *Word and object.* New edition, with a foreword by Patricia Churchland, Cambridge, Mass.: MIT Press, 2015.

Redi, L. and Shattuck-Hufnagel, S. (2001) Variation in realization of glottalization in normal speakers. *Journal of Phonetics*, *29*, 407–429.

Reinisch, E., Wozny, D.R., Mitterer, H. and Holt, L.H. (2014). Phonetic category recalibration: What are the categories? *Journal of Phonetics*, *45*, 91–105.

Remez, R.E., Fellowes, J.M. and Rubin, P.E. (1997). Talker identification based on phonetic information. *Journal of Experimental Psychology: Human Perception and Performance*, *23*, 651–666.

Remez, R.E., Van Dyk, J.L., Fellowes, J.M., and Shoretz Nagel, D. (2002). On the perception of similarity among talkers. *Barnard College Speech Perception Laboratory Technical Report*, September 2002.

Samuel, A.G. and Kraljic, T. (2009). Perceptual learning for speech. *Attention, Perception and Psychophysics*, *71*, 1207–1218.

Schacter, D.L. and Church, B.A. (1992). Auditory priming: implicit and explicit memory for words and voices. *Journal of Experimental Psychology: Learning, Memory, and Cognition*, *18*, 915–930.

Scharenborg, O., Norris, D., ten Bosch, L. and McQueen, J.M. (2005). How should a speech recognizer work? *Cognitive Science*, *29*, 867–918.

Selkirk, E.O. (1986). On derived domains in sentence phonology. *Phonology Yearbook*, *3*, 371–405.

Sheffert, S.M., Pisoni, D.B., Fellowes, J.M. and Remez, R.E. (2003). Learning to recognize talkers from natural, sinewave and reversed speech samples. *Journal of Experimental Psychology: Human Perception and Performance*, *28*, 1447–1469.

Smith, R., and Hawkins, S. (2012). Production and perception of speaker-specific phonetic detail at word boundaries. *Journal of Phonetics*, *40*, 213–233.

Smith, R., Baker, R., and Hawkins, S. (2012). Phonetic detail that distinguishes prefixed from pseudo-prefixed words. *Journal of Phonetics*, *40* (5), 689–705.

Stevens, K. N. (1989). On the quantal nature of speech. *Journal of Phonetics*, *17*, 3–45.

Theodore, R.M. and Miller, J.L. (2010). Characteristics of listener sensitivity to talker-specific phonetic detail. *The Journal of the Acoustical Society of America*, *128*, 2090–2099.

Theodore, R.M., Miller, J.L. and DeSteno, D. (2009). Individual talker differences in voice-onset-time: Contextual influences. *The Journal of the Acoustical Society of America, 125*, 3974–3982.

Trude, A., and Brown-Schmidt, S. (2012). Talker-specific perceptual adaptation during online speech perception. *Language and Cognitive Processes, 27*, 979–1001.

van den Heuvel, H., Cranen, B. and Rietveld, T. (1996). Speaker variability in the coarticulation of /a,i,u/. *Speech Communication, 18*, 113–130.

Van Heugten, M., Volkova, A., Trehub, S.E., and Schellenberg, E.G. (2014). Children's recognition of spectrally degraded cartoon voices. *Ear and Hearing, 35*, 118–125.

Weirich, M., Lancia, L., and Brunner, J. (2013). Inter-speaker articulatory variability during vowel-consonant-vowel sequences in twins and unrelated speakers. *The Journal of the Acoustical Society of America, 134*, 3766–3780.

Wickelgren, W.A. (1969). Context-sensitive coding, associative memory, and serial order in (speech) behavior. *Psychological Review, 76*, 1–15.

Frank Eisner

Donders Institute for Brain, Cognition and Behaviour, Centre for Cognition, Radboud University

Perceptual Adjustments to Speaker Variation

Abstract: Differences between speakers pose a challenge for listeners as speaker variation is among the main causes of variability in the speech signal. Listeners' ability to adapt to this variability is essential for successful comprehension. Recent research has explored how the perceptual system learns from variability by adjusting how acoustic cues are mapped onto perceptual categories. This learning can be guided by a number of different types of information, including the linguistic content of the speech, or visual cues to articulation from the speaker's face. Properties of the learning mechanism have been identified, such as the finding that perceptual adjustments can be specific to a particular speaker and are stored for later encounters of that speaker. Learning can also generalise under certain conditions, to individuals or to a group of people. Evidence from behavioural and neuroimaging research implies a top-down process, by which learning can be driven by different types of higher-level information and results in a bias at an early acoustic-phonetic processing stage. This chapter discusses how learning helps listeners to deal with speaker variation, and considers the implications of this line of research for models of speech perception.

1. Introduction

Spoken-language comprehension requires listeners to adjust to variability in the speech signal. This variation is caused by a range of factors, including differences between speakers (e.g., in the anatomy of their vocal tract or regional accent), variation within speakers (e.g., in register, speech rate, or physiological state), but also variable signal quality (e.g., because of ambient noise, or filtering through a phone connection). While the impact of this variability is often detrimental to the performance of automatic speech recognition systems (Benzeghiba et al., 2007), human listeners can normally adjust their perception quite easily. In this chapter I review an emerging body of research which aims to understand the cognitive mechanisms underlying this plasticity in the perceptual system. This work has revealed learning processes that can act fast and induce long-lasting changes in the mapping of acoustical cues onto linguistically meaningful units. Psychologists have referred to this kind of adjustment as perceptual learning in the sense of Gibson (1969), who defined it as "an increase in the ability to extract information from the environment, as a result of experience and practice with stimulation coming from it." Listeners can thus be said to become better at understanding potentially difficult speech as a result of perceptual learning.

Since inter- and intra-talker variability is naturally present in speech, the ability of the perceptual system to adjust to it is essential for speech comprehension. In many traditional accounts of speech perception, variability was regarded as a nuisance, something to be discarded or 'normalised' in the process of translating the speech signal into more abstract linguistic representations (Pisoni, 1997). Recent evidence suggests, however, that not only are listeners able to adapt dynamically to sources of variability, but that they in the process encode detailed information about those sources. This knowledge can then be useful in the future in similar listening situations. For example, being familiar with a speaker's voice makes it easier to understand that person in a noisy listening situation (Nygaard and Pisoni, 1998; Nygaard, Sommers, and Pisoni, 1994).

As perceptual learning can become effective quickly, it is amenable to being studied in a laboratory setting. Perceptual adjustments to various sources of variability have been observed after short exposure periods on the order of minutes or hours. The dependent measure in such experiments is typically a shift in perception (e.g., a shift in the location of a phoneme category boundary), or a global increase in intelligibility (e.g., being able to repeat more words correctly) following exposure (Samuel and Kraljic, 2009). Learning can thus be measured respectively at the sublexical, acoustic-phonetic level, or at the lexico-semantic level. Here I will discuss some recent studies that have used perceptual learning paradigms in order to understand basic properties of the adaptation process – when it occurs, what constrains it, how general or specific it is, and what kinds of information in the speech signal can drive it. Although these subtle changes in perception are still quite difficult to track with neuroimaging methods, there is recent evidence showing that this type of learning affects early processing stages in the auditory cortex, supporting the idea that relatively high-level sources of information can drive changes at a relatively low perceptual level. Understanding the mechanisms which enable this adaptability thus gives us a more complete picture of spoken-language processing. I will end by discussing some implications of this literature for computational and neurobiological models of speech perception.

2. Adjusting perceptual categories

There is ample evidence that listeners can adapt to a range of different types of variability in the speech signal, such as in synthetic (Fenn et al., 2003; Greenspan, Nusbaum, and Pisoni, 1988), time-compressed (Dupoux and Green, 1997) or noise-vocoded speech (Rosen et al., 1999), speech embedded in multi-speaker babble noise (Song et al., 2012), and accents (Clarke and Garrett, 2004; Weber et al., 2014). In foreign-accented speech, for example, significant processing gains begin to emerge after exposure to only a few accented sentences (Clarke and Garrett, 2004; Weber et al., 2014). These studies have typically used either an increase in intelligibility, as measured by having listeners repeat or transcribe what they heard, or an increase in processing speed, as measured by reaction times in a comprehension-based task, as the dependent variable.

A central question in the context of speaker-specific listening is whether this kind of learning, such as adapting to a foreign accent, can also generalise and aid in the comprehension of other speakers who speak with the same accent. This was investigated in a series of experiments on Chinese-accented English with American listeners by Bradlow and Bent (2008). In their study, listeners were trained to become better at understanding Chinese-accented speech coming either from only one speaker or from several different speakers. After training, generalisation of learning was tested with speech materials from an unfamiliar speaker. For listeners in both conditions, intelligibility of the accented speech increased during training. However, only after exposure to multiple speakers was there evidence of speaker-independent learning. Thus, the perceptual system seemed to treat the unfamiliar accent initially as a speaker idiosyncrasy, but was able to construct a more abstract representation of that accent after exposure to it from multiple speakers. This behaviour is adaptive in the sense that it would not be beneficial to apply learning about a speaker idiosyncrasy indiscriminately, since any given novel speaker is unlikely to have that same idiosyncrasy in their speech. It is beneficial however, to have a more abstract representation of non-standard features that apply to a larger group, because the learned representation can be applied immediately rather than having to go through the learning process over and over again for every encounter of a new speaker with that accent.

While this type of empirical research has revealed important properties of perceptual learning about speakers, measuring global comprehension by testing at the lexical level, cannot identify what exactly it is in the speech signal that listeners are adapting to, or how they do it. However, a related series of studies has investigated how perceptual learning affects processing at a sublexical level, and the mechanisms that may be driving it. These experiments used an ambiguous speech stimulus, that is, a sound that falls on the category boundary between two phonemes, as a proxy for a speaker idiosyncrasy or a feature of an accent. Learning is measured by observing relatively subtle shifts in the categorisation of such ambiguous stimuli following a period of exposure. During exposure, listeners have different types of contextual information available that can disambiguate the perception of such sounds. In fact, there are several sources of information that can drive learning, including lexical, visual, and sublexical cues, which are discussed in turn below.

A seminal study by Norris and colleagues demonstrated that listeners can use lexical knowledge of their language to guide perception of speech sounds at a sublexical level (Norris et al., 2003). For example, an ambiguous fricative that is midway between /s/ and /f/ is perceived as /s/ when placed in a context like "albatro–", but is perceived as an /f/ at the end of a word like "paragra–" (Ganong, 1980). Repeated exposure to the ambiguous sound in such lexically-biased contexts leads to a recalibration of the category boundary between /s/ and /f/ in a way that is consistent with the lexical context (see Figure 1, Eisner and McQueen, 2006): Listeners who heard the ambiguous sound in words where it replaced an /f/ subsequently categorised more sounds on an /f/-/s/ continuum as /f/, while, conversely, listeners who had heard the same ambiguous sound in /s/-biased contexts subsequently categorised more sounds as /s/ (Norris et al., 2003). A control condition, in which the same ambiguous sound was embedded in nonwords, produced no shift in categorisation responses. This pattern of results suggests that listeners use lexical information to adjust their perception of an ambiguous sound after only brief exposure to this speaker idiosyncrasy (in this case, 12 instances of the critical sound during exposure).

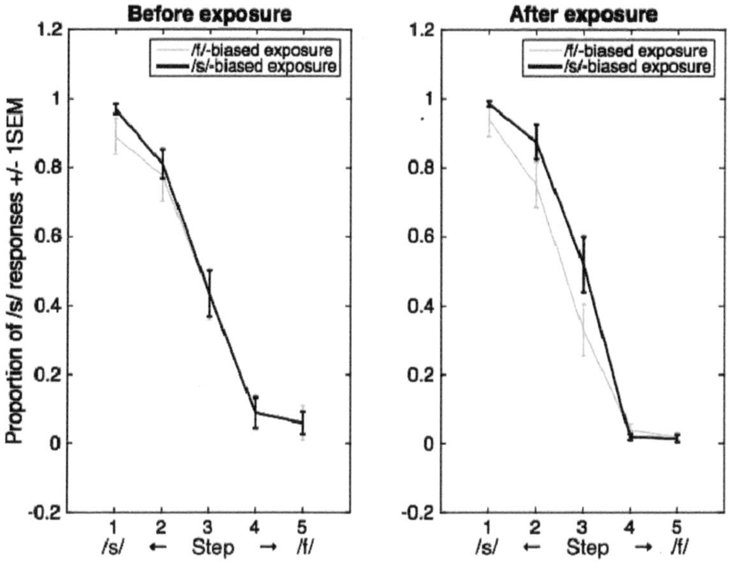

Figure 1. Perceptual learning effect in a pretest–exposure–posttest design analogous to that of Eisner and McQueen (2006; unpublished data). Two groups of listeners first categorised sounds from a 5-step /s/-/f/ continuum. Their responses were equivalent before exposure (left panel). Participants then heard the most ambiguous step 3 embedded in 2.5 minutes of continuous speech, where it replaced all /f/ sounds for one group, and all /s/ sounds for the other group. Categorisation of step 3 shifted following exposure, such that listeners with /s/-biased exposure gave more /s/ responses, and listeners with /f/-biased exposure gave more /f/ responses.

While the paradigm by Norris and colleagues is based on the lexical influence on phoneme perception (i.e., the Ganong effect; Ganong, 1980), a related paradigm is based on a similar influence from the visual domain (i.e., the McGurk effect; McGurk and MacDonald, 1976). The original McGurk effect demonstrated that auditory and visual cues are immediately integrated in perception, by showing that a video of a talker articulating the syllable /ba/ combined with a clear auditory /ga/ often results in the fused percept of /da/. A more recent study found that visual cues can also drive auditory recalibration in situations where ambiguous auditory information is disambiguated by visual information: When perceivers repeatedly hear a sound which could be either /d/ or /b/, presented together with a video of a speaker producing /d/, their phonetic category boundary shifts in a way that is consistent with the information they receive through lip-reading, and the ambiguous sound is assimilated into the /d/ category. However, when the same ambiguous sound is presented with the speaker producing /b/, the boundary shift occurs in the opposite direction (Bertelson et al., 2003; Vroomen and Baart, 2009a). Thus, listeners can use information from the visual modality to retune their perception of ambiguous speech input; in this case long-term knowledge about the co-occurrence of certain visual and acoustic cues (but also from orthographic mapping; Mitterer and McQueen, 2009).

In addition to visually- and lexically-driven recalibration, two types of sublexical information have been shown to drive similar learning effects. One is the phonotactic regularities of a language. For example, the English sequence "–rul" is phonotactically legal if the initial sound is /f/, but illegal if it was /s/. The reverse case is a sequence like "–nud", where the nonword 'snud' is consistent with English phonotactics, but 'fnud' is not. In direct analogy to lexically- and visually-driven learning, listeners can exploit these statistical regularities when the acoustic signal is ambiguous: Repeatedly hearing an ambiguous /s/–/f/ fricative in contexts like "–rul" results in a shift of the category boundary towards /f/, whereas hearing the same sound in contexts like "–nud" results in a shift to /s/ (Cutler et al., 2008). A second type of sublexically-driven adaptation is induced by contingencies between acoustic cues that make up a phonetic category, such as the multidimensional cues to the identity of stop consonants. For example, one of the main differences between /b/ and /p/ is a temporal distinction in the onset of voicing (VOT), but one of the secondary cues is the fundamental frequency (F0) of a following vowel. Because these two cues co-occur in a predictable manner (shorter VOTs occur with low F0; longer VOTs with high F0), listeners have implicit knowledge which, again, can be exploited when the speech signal is unclear: Repeated exposure to a stop with ambiguous VOT, in an F0 context which is either consistent with /b/ or /p/, will lead listeners to adjust their category boundary for /b/ and /p/ accordingly over time (Idemaru and Holt, 2011).

Sublexical category adjustments can thus be guided by various kinds of language-specific information. Research using the exposure–test paradigm to induce phonetic recalibration has revealed some fundamental properties of how listeners adjust to speaker idiosyncrasies. The learning is fast and does not require explicit attention (McQueen et al., 2006b). While listeners are not usually conscious of the shift, learning can be modulated by high-level contextual information. For example, learning is blocked when the source of the ambiguity can be attributed to a transient event, such as the speaker having a pen in her mouth, rather than an inherent characteristic of the speaker (Kraljic et al., 2008). It has been shown to remain stable for a period of up to one week (Eisner and McQueen, 2006; Witteman et al., 2015), although the effect dissipates after prolonged testing involving unambiguous sounds (van Linden and Vroomen, 2007; Vroomen and Baart, 2009b). In parallel with research on generalisation of learning about a foreign accent, several studies have investigated whether category recalibration is speaker-specific or speaker-independent, by changing the speaker between exposure and test phase. This work so far has produced mixed results, sometimes finding evidence of generalisation across speakers (Kraljic and Samuel, 2006; 2007; Reinisch and Holt, 2014) and sometimes evidence of speaker-specificity (Eisner and McQueen, 2005; Kraljic and Samuel, 2007; Reinisch et al., 2014). The divergent findings might be partly explained by considering the perceptual similarity between tokens from the exposure and test speakers (Kraljic and Samuel, 2007; Reinisch and Holt, 2014). When there is a high degree of similarity in the acoustic-phonetic properties of the critical phoneme, it appears to be more common that learning transfers from one speaker to another.

There is thus evidence from a variety of sources that speaker-specific information in the signal influences speech perception. Strikingly, there is also evidence that listeners' beliefs about who is talking are enough to have an impact on perception (Rubin, 1992). For example, the perceived ethnicity of a speaker can affect how intelligible listeners find their speech. In one study, when primed with a photo of a Chinese Canadian speaker, native listeners judged speech materials as more accented, and less intelligible, than when the same speech materials were presented without a photo. No such effect occurred when the prime was a photo of a White Canadian speaker (Babel and Russell, 2015). Effects of perceived speaker identity are not limited to global intelligibility or accentedness ratings, but have been found also at a sublexical level. Listeners take their knowledge of foreign and regional accents into account when making judgements about individual speech sounds (Hay et al., 2006; Jannedy et al., 2011; Niedzielski, 1999). For example, listeners reported hearing more raised variants of the vowel /ɪ/ in spoken sentences when primed with the written word 'Australian' than when primed with the word 'New Zealander' and hearing the same sentences (Hay et al., 2006). This pattern is in line with the typical /ɪ/ productions of talkers from Australia and New Zealand. In a recent study, we asked whether the perceived accent of a talker would also influence how likely listeners are to make a perceptual adjustment to that talker's idiosyncratic pronunciations (Eisner et al., 2013). The idiosyncrasy in this case was word-final devoicing of English stop consonants, which often occurs in learners of English whose native language is Dutch, German, or Turkish, among others. Native English listeners were exposed to Dutch-accented English which contained devoiced stop consonants at the end of words (e.g., 'seed' pronounced more like 'seat'), but not in any other positions. These listeners appeared to adjust to the devoicing by expanding their category for the voiced stop /d/, as measured immediately after exposure. The learning generalised to other positions in the word, such that words with initial voiceless stop consonants such as 'town' were acceptable instances for words that should be voiceless, such as 'down.' Interestingly, this generalisation to from word-final to word-initial position was only found with genuine Dutch-accented speech, but not in a second experiment in which the speaker was English native and purposefully mimicked the final devoicing. In that case, listeners adjusted to the devoicing, but did not generalise the learning to other positions. The perceived global accent of the speaker thus appears to constrain how listeners perceive individual speech sounds, but also the way in which they adjust to a talker idiosyncrasy.

To summarise, previously acquired knowledge about non-standard productions of a particular speaker, or a group of speakers, can affect sub-lexical processes in general and perceptual learning in particular. This ability of the system to utilise this kind of previously learned information has implications for models of speech perception.

3. Speaker idiosyncrasies in models of speech perception

3.1. Computational models

Adjusting to speaker idiosyncrasies as described above is not yet fully explained by current computational models of speech comprehension. Two broad classes of models of speech perception are distinguished on the basis of the granularity of acoustic-phonetic information as the signal is being processed from sound wave to meaning: abstractionist and episodic models. In abstractionist models such as TRACE (McClelland and Elman, 1986), the Distributed Cohort Model (Gaskell and Marslen-Wilson, 1997), or Shortlist (Norris, 1994), acoustic-phonetic detail, including information about the speaker, does not feature in the computations leading up to word recognition. These models have a layered architecture with a lexical level at the top and abstract, phoneme-like units mediating between the speech signal and the lexicon. TRACE, for example, can also in principle account for general learning effects because it has top-down connections across the system by which lexical information can modulate sublexical processing. However, because the input to those models consists of abstract units not containing fine phonetic detail, an adjustment at a sublexical processing stage would always generalise across the system, regardless of who the speaker is: There is no mechanism to incorporate prior knowledge about the speaker into the processing stream. In contrast, episodic models such as MINERVA (Goldinger, 1998) encode detailed memory traces about every spoken word they encounter, and do not feature abstract sublexical units. During word recognition, lexical candidates are activated in proportion to the similarity between the input signal and memory traces. This lack of abstraction means that fine phonetic detail remains part of the representation. Episodic models are thus able to explain speaker-specific learning effects. However, this type of model fails to account for a different finding in the literature on perceptual learning of speaker idiosyncrasies: The learning has a broad effect in the sense that it applies beyond the specific instances heard during exposure, and generalises to other words in the listener's mental lexicon (McQueen et al., 2006a), even words of other languages when spoken by the same talker (Reinisch et al., 2012). This generalisation is difficult to explain without a prelexical processing layer containing abstract representations that are connected to all entries in the lexicon (Cutler et al., 2010). In an episodic model, a learned adjustment remains specific to the exposure items, whereas in an abstractionist model, a prelexical recalibration of a phoneme contrast will affect all words in the lexicon which contain that contrast. In summary, both classes of computational model remain insufficient for explaining recent data on how listeners adjust to speech, and the evidence may point towards some kind of hybrid model. In such a model, fine phonetic detail, for example speaker-specific information, needs to be taken into account in the decoding of the speech signal. The output of these early perceptual processes might be conceived of as being probabilistic, such that the input to the word recognition system consists of phoneme likelihoods rather than strings of abstract phoneme categories (as in the revised Shortlist B model; Norris and McQueen, 2008).

3.2. Neurobiological models

The idea of an acoustic-phonetic processing system which can take into account fine phonetic detail of previously learned episodes has also received some support from neuroscience. Research in this area has identified several candidate regions in superior temporal and inferior parietal cortex (Chan et al., 2013; Obleser and Eisner, 2009; Turkeltaub and Coslett, 2010) that are engaged in aspects of processing speech at a sublexical level of analysis. Like some of the computational models, models of the neurobiology of speech perception incorporate the notion of a functional hierarchy in the processing of sound, and speech in particular. A hierarchical division of the auditory cortex underlies the processing of simple to increasingly complex sounds both in non-human primates (Kaas and Hackett, 2000; Petkov et al., 2006; Rauschecker and Tian, 2000) and in humans (e.g., Binder et al., 1997; Liebenthal et al., 2005; Obleser and Eisner, 2009; Scott and Wise, 2004). Beyond these early acoustic phonetic stages, processing streams extending in antero-ventral and postero-dorsal direction from primary auditory cortex have been identified (Hickok and Poeppel, 2007; Rauschecker and Tian, 2000; Rauschecker and Scott, 2009; Scott and Johnsrude, 2003). In the left hemisphere, the anterior stream is usually attributed with decoding linguistic meaning (Davis and Johnsrude, 2003; Hickok and Poeppel, 2007; Scott et al., 2000). In contrast, the anterior stream in the right hemisphere appears to be less sensitive to linguistic information, and more sensitive to information about speakers more generally. Studies that have investigated cortical responses to human vocal sounds in general, and to speaker variation in particular, have found activations primarily on the right (Belin and Zatorre, 2003; Belin et al., 2000; Formisano et al., 2008; Kriegstein and Giraud, 2004; Kriegstein et al., 2008; Kriegstein et al., 2003); and there is converging evidence for conspecific vocalisations in non-human primates (Petkov et al., 2008). The literature thus suggests that there are right-lateralised regions in the auditory cortex that are engaged in the processing of speaker-specific information in speech, but it is currently unclear whether these systems support speech perception, for example by making available speaker-specific information that can be integrated by an early acoustic-phonetic processing system in the left hemisphere.

Nevertheless, there is some evidence from neuroscience for this kind of modulation of early acoustic-phonetic processing. Although it did not specifically investigate speaker-specificity, a recent study by Kilian-Hütten et al. (Kilian-Hütten et al., 2011) demonstrated that early acoustic-phonetic processing is indeed affected by previously learned biases. This study found direct evidence of dynamic adjustments to a phonetic category in left auditory cortex: Using a visually-guided perceptual recalibration paradigm (Bertelson et al., 2003), regions of primary auditory cortex (specifically, Heschl's gyrus and sulcus, extending into planum temporale) could be identified whose activity pattern specifically reflected listeners' adjusted percepts after exposure, rather than simply physical properties of the stimuli. This suggests not only a bottom-up mapping of acoustical cues to perceptual categories in left auditory cortex, but it also shows that the mapping involves the integration of previously learned knowledge within the same auditory areas; in this case, coming from the visual system. Whether linguistic processing in left auditory cortex can be driven by other types of information, such as speaker-specific knowledge from the right anterior stream will be an interesting question for future empirical investigation.

4. Conclusions

Plasticity in the mapping of acoustic features to perceptual categories underlies listeners' ability to adjust rapidly to idiosyncratic properties of individual speakers. Once an adjustment has been learned, it can be used again for later encounters with a speaker. The evidence from the perceptual learning literature is compatible with a system in which such learned biases are integrated with bottom-up properties of the signal early on during processing, and suggests that the output of this system is probabilistic in nature. However, these processes cannot yet be fully accounted for by current computational and neurobiological models. Perceptual adjustments can be driven by a variety of different sources, such as visual, lexical, and sublexical – and possibly more that are yet to be identified. Studying perceptual adaptation in response to speaker variability is becoming feasible with advanced neuroimaging methods, and this promises to be a valuable tool for probing the neural underpinnings of sublexical processing and abstraction.

Acknowledgements

FE is supported by the research consortium "Language in Interaction" from the Dutch Science Foundation (NWO), and part of this work was funded by NWO grant 275-75-009 to the author. Thanks to two anonymous reviewers for their comments on an earlier version of the manuscript.

References

Babel, M., and Russell, J. (2015). Expectations and speech intelligibility. *The Journal of the Acoustical Society of America, 137*(5), 2823–2833.

Belin, P., and Zatorre, R. J. (2003). Adaptation to speaker's voice in right anterior temporal lobe. *NeuroReport, 14*, 2104–2109.

Belin, P., Zatorre, R. J., Lafaille, P., Ahad, P., and Pike, B. (2000). Voice-selective areas in human auditory cortex. *Nature, 403*, 309–312.

Benzeghiba, M., De Mori, R., Deroo, O., Dupont, S., Erbes, T., Jouvet, D., et al. (2007). Automatic speech recognition and speech variability: A review. *Speech Communication, 49*(10-11), 763–786.

Bertelson, P., Vroomen, J., and de Gelder, B. (2003). Visual recalibration of auditory speech identification: A McGurk aftereffect. *Psychological Science, 14*, 592–597.

Binder, J. R., Frost, J. A., Hammeke, T. A., Cox, R. W., Rao, S. M., and Prieto, T. (1997). Human brain language areas identified by functional magnetic resonance imaging. *Journal of Neuroscience, 17*, 353–362.

Bradlow, A. R., and Bent, T. (2008). Perceptual adaptation to non-native speech. *Cognition, 106*(2), 707–729.

Chan, A. M., Dykstra, A. R., Jayaram, V., Leonard, M. K., Travis, K. E., Gygi, B., et al. (2013). Speech-specific tuning of neurons in human superior temporal gyrus. *Cerebral Cortex. Cortex,* first published online May 16, 2013 doi:10.1093/cercor/bht127

Clarke, C. M., and Garrett, M. F. (2004). Rapid adaptation to foreign-accented English. *The Journal of the Acoustical Society of America, 116*(6), 3647–3658.

Cutler, A., Eisner, F., McQueen, J. M., and Norris, D. (2010). How abstract phonemic categories are necessary for coping with speaker-related variation. *Laboratory Phonology, 10*, 91–111.

Cutler, A., McQueen, J., Butterfield, S., and Norris, D. (2008). Prelexically-driven perceptual retuning of phoneme boundaries. In *Proceedings of Interspeech-2008*, 2056.

Davis, M. H., and Johnsrude, I. S. (2003). Hierarchical processing in spoken language comprehension. *Journal of Neuroscience, 23*(8), 3423–3431.

Dupoux, E., and Green, K. (1997). Perceptual adjustment to highly compressed speech: effects of talker and rate changes. *Journal of Experimental Psychology: Human Perception and Performance, 23*(3), 914–927.

Eisner, F., and McQueen, J. M. (2005). The specificity of perceptual learning in speech processing. *Perception & Psychophysics, 67*(2), 224–238.

Eisner, F., and McQueen, J. M. (2006). Perceptual learning in speech: Stability over time. *The Journal of the Acoustical Society of America, 119*(4), 1950–1953.

Eisner, F., Melinger, A., and Weber, A. (2013). Constraints on the transfer of perceptual learning in accented speech. *Frontiers in Psychology, 4*, 148.

Fenn, K. M., Nusbaum, H. C., and Margoliash, D. (2003). Consolidation during sleep of perceptual learning of spoken language. *Nature, 425*, 614–616.

Formisano, E., De Martino, F., Bonte, M., and Goebel, R. (2008). "Who" is saying "what?" Brain-based decoding of human voice and speech. *Science (New York, NY), 322*(5903), 970–973.

Ganong, W. F. (1980). Phonetic categorization in auditory word perception. *Journal of Experimental Psychology: Human Perception and Performance,* 6, 110–125.

Gaskell, M. G., and Marslen-Wilson, W. D. (1997). Integrating form and meaning: A distributed model of speech perception. *Language and Cognitive Processes, 12*, 613–656.

Gibson, E. J. (1969). Principles of perceptual learning and development. *Book*. Englewood Cliffs, NJ.

Goldinger, S. D. (1998). Echoes of echoes? An episodic theory of lexical access. *Psychological Review, 105*, 251–279.

Greenspan, S. L., Nusbaum, H. C., and Pisoni, D. B. (1988). Perceptual learning of synthetic speech produced by rule. *Journal of Experimental Psychology: Learning, Memory, and Cognition, 14*(3), 421–433.

Hay, J., Nolan, A., and Drager, K. (2006). From fush to feesh: Exemplar priming in speech perception. *The Linguistic Review, 23*(3), 351.

Hickok, G., and Poeppel, D. (2007). The cortical organization of speech processing. *Nature Reviews Neuroscience, 8*, 393–402.

Idemaru, K., and Holt, L. L. (2011). Word recognition reflects dimension-based statistical learning. *Journal of Experimental Psychology: Human Perception and Performance, 37*(6), 1939–1956.

Jannedy, S., Weirich, M., and Brunner, J. (2011). The effect of inferences on the perceptual categorization of Berlin German fricatives. In *Proceedings of the International Congress of Phonetic Sciences, Hong Kong*, pp. 962–965.

Kaas, J. H., and Hackett, T. A. (2000). Subdivisions of auditory cortex and processing streams in primates. *Proceedings of the National Academy of Sciences, USA, 97*(22), 11793–11799.

Kilian-Hütten, N., Valente, G., Vroomen, J., and Formisano, E. (2011). Auditory cortex encodes the perceptual interpretation of ambiguous sound. *Journal of Neuroscience, 31*(5), 1715–1720.

Kraljic, T., and Samuel, A. G. (2006). Generalization in perceptual learning for speech. *Psychonomic Bulletin and Review, 13*(2), 262–268.

Kraljic, T., and Samuel, A. G. (2007). Perceptual adjustments to multiple speakers. *Journal of Memory and Language, 56*, 1–15.

Kraljic, T., Samuel, A. G., and Brennan, S. E. (2008). First impressions and last resorts: how listeners adjust to speaker variability. *Psychological Science, 19*(4), 332–338.

Kriegstein, K. V., and Giraud, A.-L. (2004). Distinct functional substrates along the right superior temporal sulcus for the processing of voices. *NeuroImage, 22*, 948–955.

Kriegstein, K. V., Dogan, O., Grüter, M., Giraud, A.-L., Kell, C. A., Grüter, T., et al. (2008). Simulation of talking faces in the human brain improves auditory speech recognition. *Proceedings of the National Academy of Sciences of the United States of America, 105*(18), 6747–6752.

Kriegstein, K. V., Eger, E., Kleinschmidt, A., and Giraud, A. L. (2003). Modulation of neural responses to speech by directing attention to voices or verbal content. *Cognitive Brain Research, 17*(1), 48–55.

Liebenthal, E., Binder, J. R., Spitzer, S. M., Possing, E. T., and Medler, D. A. (2005). Neural substrates of phonemic perception. *Cerebral Cortex, 15*, 1621–1631.

McClelland, J. L., and Elman, J. L. (1986). The TRACE model of speech perception. *Cognitive Psychology, 18*, 1–86.

McGurk, H., and MacDonald, J. (1976). Hearing lips and seeing voices. *Nature, 264*, 746–748.

McQueen, J. M., Cutler, A., and Norris, D. (2006a). Phonological abstraction in the mental lexicon. *Cognitive Science: a Multidisciplinary Journal, 30*(6), 1113–1126.

McQueen, J. M., Norris, D., and Cutler, A. (2006b). The dynamic nature of speech perception. *Language and Speech, 49*, 101–112.

Mitterer, H., and McQueen, J. M. (2009). Foreign subtitles help but native-language subtitles harm foreign speech perception. *PloS One*, *4*(11), e7785.

Niedzielski, N. (1999). The effect of social information on the perception of sociolinguistic variables. *Journal of Language and Social Psychology, 18*(1), 62–85.

Norris, D. (1994). Shortlist: A connectionist model of continuous speech recognition. *Cognition, 52*, 189–234.

Norris, D., and McQueen, J. M. (2008). Shortlist B: A Bayesian model of continuous speech recognition. *Psychological Review, 115*, 357–395.

Norris, D., McQueen, J. M., and Cutler, A. (2003). Perceptual learning in speech. *Cognitive Psychology, 47*, 204–238.

Nygaard, L. C., and Pisoni, D. B. (1998). Talker-specific learning in speech perception. *Perception and Psychophysics, 60*, 355–376.

Nygaard, L. C., Sommers, M. S., and Pisoni, D. B. (1994). Speech perception as a talker-contingent process. *Psychological Science, 5*(1), 42–46.

Obleser, J., and Eisner, F. (2009). Pre-lexical abstraction of speech in the auditory cortex. *Trends in Cognitive Sciences, 13*(1), 14–19.

Petkov, C. I., Kayser, C., Augath, M., and Logothetis, N. K. (2006). Functional imaging reveals numerous fields in the monkey auditory cortex. *PLOS Biology, 4*, 1–14.

Petkov, C. I., Kayser, C., Steudel, T., Whittinstall, K., Augath, M., and Logothetis, N. K. (2008). A voice region in the monkey brain. *Nature Neuroscience, 11*, 367–374.

Pisoni, D. B. (1997). Some thoughts on 'normalization' in speech perception. *Collection* (pp. 9–30).

Rauschecker, J. P., and Tian, B. (2000). Mechanisms and streams for processing of "what" and "where" in auditory cortex. *Proceedings of the National Academy of Sciences of the United States of America*, 97(22), 11800–11806.

Rauschecker, J., and Scott, S. (2009). Maps and streams in the auditory cortex: nonhuman primates illuminate human speech processing. *Nature Neuroscience, 12*(6), 718–724.

Reinisch, E., and Holt, L. L. (2014). Lexically guided phonetic retuning of foreign-accented speech and its generalization. *Journal of Experimental Psychology: Human Perception and Performance, 40*(2), 539–555.

Reinisch, E., Weber, A., and Mitterer, H. (2012). Listeners retune phoneme categories across languages. *Journal of Experimental Psychology: Human Perception and Performance, 39*(1), 75–86.

Reinisch, E., Wozny, D. R., Mitterer, H., and Holt, L. L. (2014). Phonetic category recalibration: What are the categories? *Journal of Phonetics, 45*, 91–105.

Rosen, S., Faulkner, A., and Wilkinson, L. (1999). Adaptation by normal listeners to upward spectral shifts of speech: Implications for cochlear implants. *The Journal of the Acoustical Society of America, 106*(6), 3629–3636.

Rubin, D. (1992). Nonlanguage factors affecting undergraduates' judgments of nonnative English-speaking teaching assistants. *Research in Higher Education, 33*(4), 511–531.

Samuel, A. G., and Kraljic, T. (2009). Perceptual learning for speech. *Attention, Perception & Psychophysics, 71*(6), 1207–1218.

Scott, S. K., and Johnsrude, I. S. (2003). The neuroanatomical and functional organization of speech perception. *Trends in Neurosciences, 26*, 100–107.

Scott, S. K., and Wise, R. J. S. (2004). The functional neuroanatomy of prelexical processing in speech perception. *Cognition, 92*, 13–45.

Scott, S. K., Blank, C. C., Rosen, S., and Wise, R. J. S. (2000). Identification of a pathway for intelligible speech in the left temporal lobe. *Brain, 123*, 2400–2406.

Song, J. H., Skoe, E., Banai, K., and Kraus, N. (2012). Training to improve hearing speech in noise: Biological mechanisms. *Cerebral Cortex, 22*(5), 1180–1190.

Turkeltaub, P. E., and Coslett, H. B. (2010). Localization of sublexical speech perception components. *Brain and Language, 114*(1), 1–15.

van Linden, S., and Vroomen, J. (2007). Recalibration of phonetic categories by lipread speech versus lexical information. *Journal of Experimental Psychology: Human Perception and Performance, 33*(6), 1483–1494.

Vroomen, J., and Baart, M. (2009a). Phonetic recalibration only occurs in speech mode. *Cognition, 110(2)*, 254–259.

Vroomen, J., and Baart, M. (2009b). Recalibration of phonetic categories by lipread speech: measuring aftereffects after a 24-hour delay. *Language and Speech*, 52(Pt 2-3), 341–350.

Weber, A., Di Betta, A. M., and McQueen, J. M. (2014). Treack or trit: Adaptation to genuine and arbitrary foreign accents by monolingual and bilingual listeners. *Journal of Phonetics*, *46*, 34–51.

Witteman, M. J., Bardhan, N. P., Weber, A., and McQueen, J. M. (2015). Automaticity and stability of adaptation to a foreign-accented speaker. *Language and Speech 58:* 168–189.

Marieke van Heugten, Christina Bergmann and Alejandrina Cristia

Laboratoire de Sciences Cognitives et Psycholinguistique (ENS / EHESS / CNRS)
Département d'Études Cognitives, École Normale Supérieure – PSL Research University

The Effects of Talker Voice and Accent on Young Children's Speech Perception

Abstract: Within the first few years of life, children acquire many of the building blocks of their native language. This not only involves knowledge about the linguistic structure of spoken language, but also knowledge about the way in which this linguistic structure surfaces in their speech input. In this chapter, we review how infants and toddlers cope with differences between speakers and accents. Within the context of milestones in early speech perception, we examine how voice and accent characteristics are integrated during language processing, looking closely at the advantages and disadvantages of speaker and accent familiarity, surface-level deviation between two utterances, variability in the input, and prior speaker exposure. We conclude that although deviation from the child's standard can complicate speech perception early in life, young listeners can overcome these additional challenges. This suggests that early spoken language processing is flexible and adaptive to the listening situation at hand.

1. Introduction

Human communication appears to be effortless: Under optimal listening conditions we hardly experience difficulty understanding other people who speak our native language. Language comprehension is, however, far from trivial. Although theories differ in their implementation of the way in which words are accessed in the mental lexicon, it is clear that spoken language is often ambiguous in nature and thus triggers the simultaneous activation of multiple – partially overlapping – word candidates, all competing for recognition. Ultimately, in the case of successful language comprehension, one of the candidate words should be recognized as the target. How does this activation and selection mechanism work?

Answering this question is not as easy as one may think. This is partially due to the fact that speech perception is greatly complicated by the absence of a one-to-one correspondence between the surface forms of words and their underlying linguistic representation. That is, factors such as speech rate, the neighboring linguistic content, but also the speaker's voice or accent, can dramatically alter the pronunciation of words across utterances. Let us, for example, consider a female American English speaker from California and a male British English speaker from London, both producing the word *grass*. As adults, we immediately grasp that although the two word tokens differ on multiple dimensions (e. g., high-pitched Californian [ɡɹæs] versus low-pitched London [ɡɹɑːs]), they nonetheless both refer to the same underlying representation of narrow green-leafed plants commonly grown on lawns and in gardens. We also understand that both pronunciations are functionally different from phonologically closely related words such as [ɡɹəʊs], *gross*. In order to become proficient language users, children must acquire sufficient language expertise to make both inferences when they process speech. In other words, they must learn to strike a sophisticated balance between the use of linguistic and speaker-specific cues during word recognition. This chapter deals with how young children accomplish this impressive feat.

Children learn their native language with tremendous speed. By the time they reach their first birthday, most infants will have produced their first words. But even in the preceding months, children acquire numerous aspects of their native language. By six months of age, for example, they will have developed some understanding of frequently occurring words in the input directed to them (Bergelson and Swingley, 2012; Tincoff and Jusczyk, 1999, 2012) and they will recognize these words when spoken by a speaker they have never heard before (Bergelson and Swingley, 2013; Mandel et al., 1995; Tincoff and Jusczyk, 1999, 2012). This suggests that children's lexicons develop early in life and that even the initial word representations are sufficiently robust to deal with the variability between speakers.

This does not, however, mean that young children completely disregard speaker-specific information. In fact, much like adults, children have been shown to process speaker information to engage in non-linguistic tasks. For example, a mounting body of work shows that young children's social preferences can be greatly influenced by accent information. That is, by five to six months of age, infants prefer to look at a speaker speaking in their own native accent over a speaker speaking in a foreign accent (Kinzler et al., 2007). This early preference for native-accented speakers develops into greater trust in native speakers compared to accented speakers during the preschool period (Kinzler et al., 2011). In fact, a speaker's accent is one of the core principles children use when evaluating others. It is even more prevalent than other, perhaps visually more salient characteristics such as a person's facial morphology (Kinzler et al., 2009). This suggests that throughout early childhood, children are sensitive to and make use of the speaker-specific cues present during oral communication.

These two lines of research, showing that infants can access both the linguistic and the non-linguistic information embedded in the speech stream, suggest that in principle, children are well-equipped to take into account both types of cues. How do these cues interact during online language comprehension? Although the two types of cues originate from the same acoustic signal, it is possible that children process them separately, and that integration only takes place off-line, once each stream of information has been attended to individually. Alternatively, children may readily incorporate speaker specificities during speech perception, just like adults update their expectations about the speaker's linguistic system online (Dahan et al., 2008; Trude and Brown-Schmidt, 2012; see Cristia et al., 2012 for an overview). Distinguishing between these two possibilities has both theoretical and practical implications. Theoretically, understanding how infants contend with speaker differences allows us to establish a more complete picture of the mechanisms underlying speech perception at a young age. On a more applied level, knowing *when* typically-developing infants experience difficulty recognizing words enables us to develop strategies to overcome such difficulties. This could be particularly useful for settings in which young children encounter many different speakers (e.g., daycare or preschool). In addition, knowledge regarding children's incorporation of speaker differences could help with developing ways to identify language difficulties in children early in life.

In recent years, developmental research examining the effects of speaker variation on speech perception early in life has started to increase. In the remainder of this chapter, we will consider the ways in which infants, toddlers, and young children cope with speaker variation during language processing, in order to address how this indexical information affects linguistic processing. For the purposes of this chapter, we consider effects of voice variability to be due to differences in the physical characteristics of speakers. This includes changes in pitch, voice clarity, and resonance from one speaker to the other. By contrast, accent variability involves changes due to differences in the phonological system across speakers of the same language who often grew up in different regions. This involves, among others, shifts in the realization of certain speech sounds and differences in intonation patterns.

To examine how speaker variation affects early spoken language processing, it is important to understand the basic research conducted in the field of infant speech perception. In Section 2, we therefore first explain the early milestones of infants' linguistic processing. Readers who are unfamiliar with this research can read this section to gain an overview of the main benchmarks of spoken language acquisition during the early years, and read through the brief explanations of procedures used for this type of research in the Appendix, whereas those who are familiar with the topic of early language development may want to continue directly to Section 3. The subsequent four sections provide an overview of empirical results testing the effects of speaker differences and variability in young children. We address four main questions:

- Section 3: What are the advantages (if any) of familiarity with a voice or accent when processing spoken language?
- Section 4: What are the effects of deviations in voice and accent, from learning to recognition, during language processing?
- Section 5: How does variability between speakers' voices and accents affect spoken language processing?
- Section 6: How can prior exposure to accented speakers help with processing accented speech?

In Section 7, we conclude by integrating these different lines of research and discussing the theoretical implications of this work.

2. A primer on infant speech processing

In the past 40 years, research in the field of infant speech perception has provided us with a refined understanding of how children take their first steps in learning their native language. In this section, we summarize the salient developmental results (see Figure 1). This allows us to establish a rough timeline of infants' discovery of linguistic structure. The experimental research discussed here can be considered as the groundwork for testing the role of voice and accent variability in subsequent sections. Note that the Appendix contains an overview of many of the behavioral procedures used in infant speech perception research. When we describe work using one of the outlined procedures, asterisks are used to indicate that more detailed information regarding this paradigm is available in the Appendix.

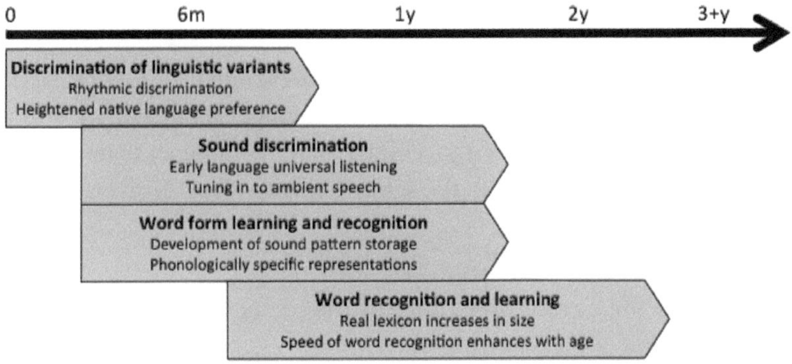

Figure 1: Infants' advances in language acquisition over the first three years of life are evident across a range of experimental tasks that have focused on certain age ranges. Presumably, development continues beyond those periods. Each of the arrows encompasses the age range typically tested in these domains, and indicates the main benchmarks infants achieve within a given area.

2.1. Discrimination and preference among languages

Languages can vary greatly in terms of their phonology. This variation not only involves differences in the use of specific sounds, but also includes differences in syllable complexity and stress. While in some languages, such as Japanese, consonants and vowels tend to alternate, other languages, such as Russian, allow for more complex syllables often containing multiple consonants. Similarly, the temporal organization of syllables within utterances differs across languages, with some languages being described as stress-timed and others as syllable- or mora-timed. Cross-linguistic research has shown that there are acoustic correlates of such differences in phonological structure present in spoken language (Ramus et al., 1999). What does this mean for young children acquiring language? Are they able to use such surface characteristics to distinguish between languages?

Studies using **Habituation*** and **Preference* Procedures** have revealed that as early as birth, infants can discriminate between pairs of languages from different rhythmic classes (Nazzi et al., 1998), and show a preference for their native language (e.g., Mehler et al., 1988; Moon et al., 1993). This pattern of results is also observed when the speech samples are low-pass filtered, but not when they are played backwards (Mehler et al., 1988). Thus, infants' language differentiation does not appear to be based on global spectral properties, such as differences in pitch or pauses, but rather seems to be based on prosodic differences between the languages. Over time, these early language discrimination abilities are further enhanced, such that by approximately five months of age, children can distinguish their native language from another language in the same rhythmic class (Nazzi et al., 2000).

2.2. Sound discrimination

In addition to the coarse phonological differences described above, languages also differ with regard to the specific speech sounds they employ. Using a variety of paradigms (such as the **Habituation Procedure*** and the **Conditioned Head Turn Procedure***), a large body of work has examined when children tune to the sound inventory of their native language. This has typically been tested through children's abilities to discriminate specific speech sounds that either do or do not occur in the ambient language. If children's sound processing is mature, they should discriminate native-language contrasts without any problems, but should – like adults – generally discriminate non-native contrasts less well.

Studies examining sound discrimination show that infants start life with the ability to discriminate most of the linguistically relevant speech sounds employed in languages throughout the world. For instance, even though English does not have the voiceless unaspirated retroflex vs. dental contrast (i.e. /ʈ/ vs. /t̪/ are contrastive in other languages, such as Hindi), English-learning 6-month-olds can discriminate these two sounds (Werker and Tees, 1984). However, with more exposure to the ambient language, infants tune in to the specific phoneme contrasts relevant to their native language. This means that they not only improve their ability to discriminate native-language sounds (e.g., Kuhl et al., 2006), but also tend to lose their ability to tell apart contrasts that are not found in their native language (although there are some salient exceptions to the general decline for non-native contrasts: Best et al., 1988; Best and McRoberts, 2003). This is not to say that learners become completely insensitive to variation occurring within a native sound category. On the contrary, certain tasks reveal that toddlers can detect within-category subphonemic variation (McMurray and Aslin, 2005). This sensitivity is potentially helpful for speaker- or accent-adaptation when individual speakers differ systematically from one another at the level of subphonemic detail.

2.3. Word form learning and recognition

During the first year of life, infants not only tune in to the sound inventory of their native language, they also start learning the word forms (i.e. the sound patterns of words) that occur frequently in their input. Work using the **Word Segmentation Procedure*** reveals that infants recognize a familiarized word form embedded in fluent speech as early as six months, depending on the infant's native language, the position of the word in a sentence, and the phonological form of the target word used (Bortfeld et al., 2005; Johnson et al., 2014; see also Bosch et al., 2013, for a discussion). In the following months, this ability stabilizes (Jusczyk and Aslin, 1995; Jusczyk et al., 1999) and by eight months of age, children store long-term representations of familiarized words that are phonemically specific (Jusczyk and Hohne, 1997). This suggests that early in life, children possess the ability to encode and store (some of) the word forms they hear in the speech stream around them.

How does this ability to represent word forms help children's processing of words that occur frequently in their real-world input? Current research using the **Frequent Word Form Procedure*** suggests that as early as five months of age children prefer to listen to their own name over a matched foil (Mandel et al., 1995), and towards the end of the first year of life they have learned many other high-frequency word forms (Hallé and De Boysson-Bardies, 1994; Swingley, 2005; Vihman et al., 2004). However, changes in the initial consonant of the frequent word form cause English-learning children to stop recognizing these items (Swingley, 2005; Vihman et al., 2004). This implies an early sensitivity to the phonemic representations of words.

2.4. Word recognition and learning

The size and content of infants' receptive lexicon is a topic of much recent work, relying mostly on measures that integrate auditory and visual information. For example, studies using the **Intermodal Preferential Looking Procedure*** have shown that infants as young as 6 months of age recognize some common nouns (Bergelson and Swingley, 2012; Tincoff and Jusczyk, 1999, 2012), although there are clear increases in both accuracy and response speed with age (Bergelson and Swingley, 2012; Fernald et al., 1998). If the word label is mispronounced, infants take longer to fixate on the correct image and show weaker preferences for that image (e.g., Mani and Plunkett, 2007, 2008; Swingley, 2009; Swingley and Aslin, 2002). These additional processing costs are proportional to the phonological distance between the mispronunciation and the target word (e.g., upon hearing *voggie* toddlers are less likely to lead to fixate on an image of a dog than upon hearing *toggie*; Mani and Plunkett, 2011; White and Morgan, 2008).

Other work has assessed toddlers' word learning. In one type of task (the **Switch Task***), 14-month-olds succeed at mapping novel labels onto novel objects when presented with pairs of words with little overlap (such as *lif* and *neem*; Stager and Werker, 1997), but not with pairs where only one segment mismatches (such as *bin* and *din*). Such minimal pairs are only learnable in this task by 17 to 20 months (Werker et al., 2002). Fourteen-month-olds' performance with minimal pairs, however, can be boosted by reducing task demands (such as using familiar words, referential cues, or presenting words in a sentential context rather than in isolation; Fennell and Waxman, 2010; Fennell and Werker, 2003; Yoshida et al., 2009).

2.5. Integrating speaker information

Before proceeding to the next section, we would like to point out that methods such as those described above not only enable us to study the acquisition of linguistic cues, but also make it possible to examine how children combine these cues with speaker-specific information during speech perception. Sound discrimination tasks, for example, can be used to assess children's reliance on surface-level aspects of the sounds by testing children's ability to generalize across sounds produced by different speakers. Similarly, word (form) recognition studies allow researchers to test children's reliance on speaker cues by measuring infants' recognition of frequent word forms spoken by an atypical or an accented speaker. And finally, in word learning tasks, experimenters can manipulate the familiarity of the speaker and the accent during the training and/or test phase to assess the role of speaker-specific information on lexical processing.

3. Effects of familiarity with the speaker's voice and/or accent

Infants' main source of language input comes from their primary caregivers. Starting approximately three months before birth, fetuses begin to perceive sensory stimulation. In the auditory domain, the mother's voice is one of the most salient contributors to prenatal sensory learning. As a result, the maternal voice has a privileged status. For example, shortly after birth, babies prefer to listen to their mother as compared to an unfamiliar female speaker (DeCasper and Fifer, 1980; Hepper et al., 1993; Mehler et al., 1978). In the following months, hearing the mother's voice leads to distinct neural activation compared to hearing an unknown speaker, as measured with Near Infrared Spectroscopy (Naoi et al., 2012), functional Magnetic Resonance Imaging (Dehaene-Lambertz et al., 2010), as well as Electroencephalography (Purhonen et al., 2004). This special role of the maternal voice has been observed across multiple languages and it remains present throughout the first year of life (see Chapter 5 in Kreiman and Sidtis, 2011, for an overview).

Since the mother's voice is so special during infancy, one may wonder whether and how it affects early speech perception. Researchers have started to investigate the possible interaction between the mother's voice and linguistic processing in young children. For example, Barker and Newman (2004) examined whether the mother's voice can help infants segregate and encode speech under challenging listening conditions. In their word segmentation experiment, 7.5-month-olds were familiarized with two word forms that were both produced either by their own mother or by an unfamiliar female talker. These familiarization words were presented simultaneously with a distracter stimulus (a second unfamiliar female speaker reading a scientific article). While infants typically succeed in this task at this age under relatively advantageous listening conditions (Jusczyk and Aslin, 1995; Newman and Jusczyk, 1996), infants who heard the words produced by an unfamiliar speaker failed to recognize the word forms at test. By contrast, those who heard the words in their own mother's voice did recognize the trained words in the subsequent test phase. Thus, in cases of adverse listening conditions void of visual, lexical, and spatial context, familiar voices may be particularly beneficial for speech segregation (see Bergmann et al., 2015 for a discussion).

An advantage for maternal language processing is also observed in word recognition work. Specifically, in a recent study, 9-month-old infants' ability to map a label onto a referent was examined using electroencephalography. Children were presented with the name of a familiar object (e.g., *duck*), followed by a visual presentation of either a matching or a mismatching object (e.g., *duck* or *book*). In the case of a mismatching object, children displayed neural signatures indicating the detection of an incongruity, but this was only observed when it was their own mother who named the objects. When the experimenter (mis-)labeled the same objects, the mismatch went unnoticed (Parise and Csibra, 2012). This makes infants' interactions with their own mother potentially more fruitful than their interactions with strangers. Note, however, that parents in this study were allowed to gesture and speak in the way they typically speak with their children, so this advantage of speaker familiarity may be due to factors other than familiarity with the mother's voice alone. Also note that both studies providing evidence for the benefits of the maternal voice during language processing have presented children with relatively difficult listening conditions (either due to having a same-gender individual speak in the background or to the asynchronous presentation of object and label) and that studies in which these challenges are reduced do not always observe such advantages (Bergelson and Swingley, 2013; Van Heugten and Johnson, 2012). It is thus plausible that the mother's voice may be particularly advantageous for situations where the processing demands are high. Indeed, under more optimal conditions, children start recognizing words produced by unfamiliar speakers of their native language from around the 6-months mark (Bergelson and Swingley, 2013; Tincoff and Jusczyk, 1999, 2012), suggesting a child's developing lexicon has the potential to be generalizable to novel speakers and novel situations. This can be very helpful when they encounter speakers they have never heard before.

So far, when discussing children's ability to understand unfamiliar speakers, we have assumed that these individuals pronounce words in approximately the same fashion as the children's parents. However, in today's linguistically diverse world, that assumption is not always a valid one. Many people live in environments where their language background does not match with that of the local community. At some point, infants will thus encounter speakers with different accents. How would children cope with such accent deviation? Do they hear the differences between accents? And if so, would they be able to understand speakers who have an unfamiliar accent?

As discussed in Section 2.1, 5-month-old infants possess the ability to differentiate between their native language and an unfamiliar language, even when that language belongs to the same rhythmic class (Nazzi et al., 2000). Will they extend this ability to the potentially more subtle differences between accents of the same language? Research using the habituation paradigm with both American and British English-learning children has revealed that although 5-month-olds are unable to discriminate between two unfamiliar accents of their native language (Butler et al., 2011), they can discriminate their own native accent from an unfamiliar accent (Butler et al., 2011; Nazzi et al., 2000). Moreover, around the same time, infants exhibit a preference for their own native accent over a completely unfamiliar accent (although their preference among their native and a more familiar accent has dissolved around this age; Kitamura et al., 2013).

Since children are sensitive to between-accent differences, one may wonder how this affects their recognition of words produced in an unfamiliar accent. Children growing up in Australia, for example, are used to hearing Australian English, whereas children growing up in Canada are more accustomed to Canadian English. It therefore stands to reason that different accents be processed differently depending on the accent background of the listener and that early word comprehension is optimized to the local accent. But can children cope with unfamiliar accents at all? To examine this question, studies have built on the finding that children prefer to listen to lists of known words over lists of unknown words (Hallé and De Boysson-Bardies, 1994; Swingley, 2005; Vihman et al., 2004). If children recognize accented pronunciations of words, such a preference pattern should emerge regardless of whether the word lists are presented in their native or in an unfamiliar accent. It is not until the second half of their second year of life, however, that children display a preference for known over unknown words in an unfamiliar accent. That is, while American English-learning 19-month-olds display a known word preference both when the speaker is American-accented and when the speaker is Jamaican-accented, 15-month-olds fail to differentiate between the known and unknown words in a Jamaican accent (Best et al., 2009; see also Van Heugten and Johnson, 2014 for similar results with Canadian children listening to Australian-accented words). In addition, although both groups display successful word identification in their native accent, 19- but not 15-month-olds identify the referent of words produced in an unfamiliar accent (Mulak et al., 2013). The exact age at which this change occurs is, however, somewhat variable across tasks and accents (see Cristia et al., 2012; Mulak and Best, 2013 for overviews), with some work pointing towards a change around 20 months of age (Best et al., 2009; Mulak et al., 2013; Van Heugten and Johnson, 2014), and other work suggesting that the ability to recognize words across accents may not evolve until later (Floccia et al., 2012; Van Heugten et al., 2015). It is thus likely that in the months preceding their second birthday, infants enter a transition period where their success in these tasks is dependent on both their linguistic maturity, potentially measured by their vocabulary size (Mulak et al., 2013; Van Heugten et al., 2015) and task demands.

4. Effects of deviation in speakers' voices and accents

The previous section dealt with the effects of infants' familiarity with a given voice and a given accent. We have seen that listening to the maternal voice (rather than an unknown voice) can have processing advantages for language comprehension. We have also presented evidence suggesting that listening to a native accented speaker (rather than listening to someone with an unfamiliar accent) can be beneficial for word recognition. We now turn to the effects of what we call *deviation,* namely the presence of discrepancy in the speaker or accent between an initial learning phase and a later test phase. Please note that we wish to keep this notion strictly distinct from that of *variability*, which involves the presence of multiple speakers and accents during the initial learning phase, and to which we will turn in the next section.

Research examining children's ability to cope with differences in the speaker's voice and affect has suggested this type of speaker-related deviation may at first be challenging. That is, even though young children have no problem recognizing word forms after only limited exposure to these items when the speaker remains unchanged (e.g., Jusczyk and Aslin, 1995) or when the speaker changes to a similar-sounding speaker (Houston and Jusczyk, 2000), they do appear to initially experience greater difficulty recognizing word forms when the speaker's voice at test is clearly different from that during familiarization (Houston and Jusczyk, 2000; Singh et al., 2004). By 7.5 months of age, for example, children familiarized with word forms such as *dog* or *feet* in a female voice later recognize these words when they are spoken by another female speaker, but not when these words are subsequently spoken by a very distinct male speaker. Only a few months later, when the child is around nine months of age, such difficulties related to voice deviation have mostly disappeared (Houston and Jusczyk, 2000). Difficulties due to accent changes are somewhat more persistent. Specifically, only by 12–13 months of age will infants generalize familiarized word forms from one accent to the other (Schmale et al., 2010; Schmale and Seidl, 2009). This decline in reliance on the exact accent-induced phonetic detail thus appears to lag a few months behind the development to learn to better cope with voice (or affect) deviation.

The finding that children are able to contend with voices before they are able to contend with accents raises an important question that we have not touched on so far. In particular, one may wonder whether children's initial difficulty to cope with voices and accents is proportional to the distance between familiarization and test items. It could, for example, be possible that children are hindered more by accent than by voice deviation simply because accents may affect the relevant acoustic-phonetic cues that children use to recognize words to a greater extent than voices do. The previously described studies suggest that while dissimilarity among voices predicts difficulty (generalization across similar voices occurs earlier than generalization across dissimilar voices), the picture is more complex when it comes to dissimilarity among accents. In word segmentation tasks, children's abilities to generalize accents emerges around the same time regardless of whether a distinct Spanish accent (Schmale and Seidl, 2009) or a much closer Canadian accent (Schmale et al., 2010) is used as the deviating accent for learners of North Midland American English. By contrast, amount of acoustic-phonetic mismatch may be more important for early word recognition. For example, while 15-month-olds have been found to reliably learn minimal pairs such as *deet* and *dit* in a word learning task (e.g., Curtin et al., 2009), success at this task only holds when the vowels of these two words are acoustically distinct in the speaker's accent (Escudero et al., 2014). When the vowels differ less on the acoustically relevant dimension, the two words are not reliably distinguished at test (Curtin et al., 2009; Escudero et al., 2014), likely because learning minimal pairs differing in just vowel quality can be challenging for young children (e.g., Nazzi, 2005; Havy and Nazzi, 2009). Thus, the generalization cost as a function of the strength of acoustic-phonetic deviation is clearly a matter for further work.

Of course, the findings that acoustic-phonetic deviation in the pronunciation of linguistic material can be challenging for infants in certain tasks does not imply that children cannot deal with any form of deviation early in life. While deviation may make linguistic processing more effortful, there is also evidence that children possess the basic capacity to deal with speaker differences early in life, both at the level of word forms (Johnson et al., 2014; Van Heugten and Johnson, 2012) and at the level of speech sounds (Kuhl, 1979, 1983). When presented with word forms in sentences, for example, rather than with isolated words in the initial familiarization phase – more similar to the way in which speech is typically heard outside the lab – children do recognize word forms in a male voice even if they had only heard them in a female voice prior to test (Van Heugten and Johnson, 2012). This suggests that while acoustic deviation can complicate word recognition, young children are, at least to some extent, equipped to deal with this challenge from early on.

5. Effects of variation in speakers' voices and accents

If acoustic *deviation* negatively affects the recognition of what is learned, does this imply that *variation* also impedes the learning process? This does not necessarily have to be the case. It seems plausible that, in contrast to what happens when learners have to generalize one speaker's pronunciation of linguistic units to a novel speaker, hearing multiple distinct speakers may help listeners construct the invariant structure (i. e. what remains the same across speakers and utterances), and would, as such, not hinder learning. If this were true, then we should observe an asymmetry where speech processing difficulties associated with multiple voices may be restricted to deviation and will not be observed for variability. Evidence for this view has been found at different levels of processing. That is, infants are able to successfully build and access linguistic representations despite (or perhaps by virtue of) variability. At the sound level, infants maintain their ability to discriminate phonemic contrasts in the face of speaker variability (Jusczyk et al., 1992; Kuhl, 1979). Similarly, word form encoding remains robust when the speaker varies during the initial learning phase (Houston, 1999; also see Singh, 2008 for similar results with affective variation). In addition to evidence for the idea that variability may not harm linguistic processing and encoding, *positive* effects of variability are observed in phonotactic learning studies: Infants presented with an artificial sound pattern grammar, in which plosive consonants are followed by lax vowels whereas fricative consonants are followed by tense vowels, are better able to learn these rules when this made-up language is uttered by multiple speakers as opposed to when it is uttered by a single speaker (Seidl, Onishi, and Cristia, 2014). Such facilitative effects are present from very early on as infants in this study benefitted from hearing multiple voices as early as four months of age. This demonstrates that voice variability can help shape the phonological patterns in the native language early on in the course of language development.

The advantage of variability can also be observed at the word level. Previous work has revealed that 14-month-old children experience difficulty learning to map two phonologically similar words (e. g., *bin* and *din*) onto different visual items, even though they successfully learned to map two phonologically unrelated words (e. g., *lif* and *neem*; Stager and Werker, 1997) onto different items. In that study, however, infants only heard a single speaker pronounce the words. To examine whether variability could help children learn phonologically similar words, researchers increased talker variation by introducing multiple voices (Rost and McMurray, 2009). When only one speaker utters a single token of each word, infants appear to conflate /buk/ and /tuk/ tokens and display no evidence of learning the mapping between word form and referent. However, when many different speakers provide the learning material, infants successfully distinguish between the two very similar forms at test. Follow-up work has furthermore revealed that it was likely the acoustic variability in linguistically irrelevant dimensions rather than the variability in the realization of the contrastive phonemes (i. e. voice onset time) that may have driven this boost in performance (Galle et al., 2015; Rost and McMurray, 2010).

Taken together, these findings reveal that exposure to speaker variability can be helpful for learning sounds and words during infancy, at least in a laboratory setting. Whether being exposed to variation in accents in everyday life can be useful in a similar fashion has not yet been examined with young children (see Levi, 2015, however for work with school-age children). A recent study revealing greater sensitivity to phonemic detail in monolingual children who hear a single accent in their language input as compared to their age-matched monolingual peers with routine exposure to multiple accents in the home environment may, however, suggest that daily exposure to accent variability could lead to less precise representations (Durrant et al., 2015; though see Van der Feest and Johnson, in press, for evidence suggesting that children with mixed accent input may simply be more flexible, rather than less precise, in their signal-to-word mapping strategies). If such results of accent variability are also observed when phonological detail is potentially more important (i. e. in cases where two phonologically similar words are learned), this could indicate that exposure to large variability might induce greater tolerance of deviation rather than greater attention to phonetic detail. Independent of the outcome of such a test case, however, the findings to date demonstrate that non-linguistic factors can alter infants' linguistic performance. This suggests that linguistic and non-linguistic information are rapidly integrated during language processing early in life.

6. Effects of brief and long-term exposure to accents

As reviewed above, understanding speech produced by someone with an unfamiliar accent is more challenging than understanding speech produced by a speaker of the listener's own accent. This holds both for children and for adults, although adults have been shown to readily adapt to unfamiliar pronunciations of words after some experience with the accent at hand (Bradlow and Bent, 2008; Clarke and Garrett, 2004; Dahan et al., 2008; Floccia, et al., 2006; Maye et al., 2008). Would brief exposure to an accented speaker also enhance children's ability to contend with accents? Studies exploring children's abilities to cope with unfamiliar accents have recently begun to look at the effects of brief exposure to a speaker. In these studies, children are first presented with a sample spoken by an accented speaker. This allows them to build a representation of the accent that can be used to understand similarly-accented input in the future. Following this initial exposure phase, children are tested on their recognition of familiar words in the exposure accent. In a first study investigating this issue, White and Aslin (2011) tested 19-month-olds on a variant of English that involved a single segment change, where low mid-front vowels were raised (leading to *dog* being pronounced as *dag*, for instance). Such exposure changed children's perception of words, such that children who had previously heard the speaker produce *dog* like *dag*, later recognized the same speaker's *battle* as *bottle* (even though they had never heard the speaker pronounce *bottle/battle* before). By contrast, children without exposure to the change did not recognize the shifted variants, and neither group tolerated *bittle* as an instance of *bottle*. This suggests that toddlers' word recognition abilities are sufficiently flexible to deal with speaker-specific differences in pronunciation, without them being too broad to accept any deviation from the native-accented form.

Although segment shifts may play a prominent role in distinguishing certain North-American English accents, dialectal differences can be much greater. Consider, for example, North-American-, Australian-, Jamaican-, Scottish-, and Spanish-accented English. Listening to only a short excerpt in each of these accents quickly reveals that they differ on more than just a single dimension. This greater deviation potentially makes accommodation harder. To examine whether children can also accommodate more distinct accents after gaining experience with that accent, a recent study tested Canadian English learners on their recognition of known words in an unfamiliar Australian accent. Without any exposure prior to test, children do not recognize the Australian-accented words until around their second birthday (Van Heugten and Johnson, 2014; Van Heugten et al., 2015). After exposure to the Australian English speaker reading a familiar story, however, 15-month-olds did recognize the Australian-accented test words (Van Heugten and Johnson, 2014). This suggests that brief exposure to the speaker may be beneficial for the recognition of familiar words in accents that are phonetically dissimilar from the child's own accent. Similarly, in a word learning study, North-American English-learning 2-year-olds were taught a novel word by a speaker of their own accent following brief exposure to either Spanish-accented speakers or to speakers of their own native accent. All children were subsequently tested on their recognition of the newly learned word spoken in a Spanish accent. Only the group previously exposed to Spanish-accented speech succeeded (Schmale, Cristia, and Seidl, 2012; see Schmale, et al., 2015 for benefits of exposure to variability more generally). This speaks to the continued use of accent experience throughout toddlerhood, at least when listening conditions are sufficiently challenging. This benefit of accent exposure is further exemplified by findings that routine exposure to a minority accent at home enables children to acquire phonological contrasts of that minority accent that do not surface in the regionally dominant accent. The contrast can then be flexibly used where necessary during online language processing depending on the speaker at hand (Van der Feest and Johnson, in press). Future work is necessary to examine how generalizable such adaptation is. Will exposure to a speaker in a given accent also allow children to better understand another speaker of that accent, perhaps even speakers of closely related accents?

Note that the finding that speaker exposure can help children contextualize their input does not mean that *any* form of experience with the speaker's accent *always* enables children to accommodate that accent. Neither short-term nor life-long exposure prevents young language listeners from experiencing difficulty understanding accented speakers in all situations. Routine exposure to a certain accent feature through one of the parents, for example, may not always be sufficient for the child to recognize words pronounced with such features in the lab. That is, 20-month-old children growing up in a rhotic accent area in the UK (where /r/ is generally preserved in all positions), but who are exposed to a non-rhotic accent (where /r/ tends to be unpronounced in postvocalic position) at home through at least one parent, experienced difficulty recognizing words in which the /r/ is not produced (Floccia et al., 2012). Prior accent experience may also be less beneficial in situations where, in the absence of a mature vocabulary, the accented pronunciations of words cannot be easily mapped onto their corresponding native-accented word forms (Van Heugten and Johnson, 2014) or when children's ability to cope with accent variability on the fly has become sufficiently robust to contend with accented speech even in the absence of exposure (Van Heugten et al., 2015). Future research will have to examine the exact conditions necessary for children to make use of speaker experience and how this relates to understanding accented speakers in the real world.

7. Conclusions

The speech signal is highly complex: In addition to linguistic data, it also conveys speaker-related information, signaling factors such as the speaker's age, sex, and regional origin. To efficiently process spoken language, listeners need to take into account these indexical factors. In the developmental literature, speaker variation has most frequently been studied using voice quality, likely because effects of voice familiarity have been observed so shortly after birth (DeCasper and Fifer, 1980; Hepper et al., 1993; Mehler et al., 1978), and because this research started to emerge before much was known about how infants perceive spoken language. In recent years, research on the integration of voice information during speech perception has been complemented by research examining the consequences of hearing speech produced by speakers of unfamiliar accents. With increasing globalization, a growing number of people move to new areas where the language background differs from what they are used to. Speakers may sound accented to members of their new community, either because of differences in the ways words are pronounced across regions or because their first language affects the pronunciation of words in their second language. In addition, global media has increased the potential for exposure to accents from different regions in the world. In this chapter, we have described how infants, toddlers, and young children cope with such variation in voice and accent during language processing.

Although effects of voice and accent could both be captured under the umbrella term "speaker-related differences", the two types of variation may in fact be different in nature. Specifically, differences in surface form due to voice quality may be considered to be acoustic, whereas differences in surface form due to accents may be thought of as being phonetic or phonological (though, of course, to examine cross-accent differences, voice differences are typically conflated with accent changes). In addition, the amount of exposure to voice variation can differ dramatically from the amount of exposure to accent variability, at least for monolingual children growing up in households in which both parents originate from the same region. Nonetheless, many of the effects of voice and accents on speech perception are convergent. For both voices and accents, listening to what is familiar has advantages for language processing (although children learn to cope with unfamiliar voices long before they learn to cope with unfamiliar accents). When learning new word forms, deviation (i. e., hearing a word spoken in a new voice or accent) furthermore tends to increase difficulty, regardless of whether the differences are due to voices or accents. Variation (i. e., hearing the same word uttered by multiple speakers), by contrast, is useful for word learning, and prior knowledge of how a speaker pronounces sounds can be helpful when contending with unfamiliar accents.

Studies examining the effects of voice and accent on infants' linguistic processing have important implications for theories of early speech perception. On the one hand, the surface form of words has been shown to play an essential role during early language processing. Infants and young children appear to be better able to recognize words and word forms when the acoustic-phonetic characteristics resemble those of previously heard instances (Best et al., 2009; Houston and Jusczyk, 2000; Mulak et al., 2013; Schmale et al., 2010; Schmale et al., 2011; Schmale and Seidl, 2009; Singh et al., 2004). This may be indicative of an exemplar-based storage system of words early in life, where speaker information is retained in the mental lexicon (Goldinger, 1996, 1998). On the other hand, young children overcome difficulties due to speaker-related discrepancies after only brief exposure to the speaker (Schmale et al., 2012; Van Heugten and Johnson, 2012, 2014; White and Aslin, 2011). This would imply that successful word recognition is not only dependent on the amount of acoustic-phonetic overlap between word tokens, but also on children's opportunity to adapt to the speaker. Moreover, this enhanced ability to recognize accented words following brief accent experience generalizes to words that have not been previously heard in the unfamiliar accent, suggesting that exposure allows children to learn the phonetic-to-phonemic mappings. Despite the emphasis on episodic storage in current models of infant speech perception (such as WRAPSA and PRIMIR; Jusczyk, 1997; Werker and Curtin, 2005, respectively), abstraction processes evidently play a significant role during word recognition. Thus, even at the early stages of spoken language processing, word representations contain an abstract component. Of course, this does not rule out the possibility that early word representations also contain exemplar information. In fact, research on adult speech perception is increasingly turning to hybrid models of spoken language processing that incorporate both exemplar theory and abstraction (e.g., Goldinger, 2007; Luce and McLellan, 2005; Pierrehumbert, 2006). In the future, this combination of episodic and abstract information in the storage of word representations should be implemented in models of infant language comprehension.

Taken together, the research on early speech perception outlined in this chapter reveals that processing spoken language that deviates from the typical language input (in terms of the speaker's voice or accent) is undeniably much more complex than processing familiar voices and accents. Nonetheless, infants and toddlers are surprisingly capable to contend with voice and accent deviation. With only brief speaker exposure, for example, children can overcome the additional processing costs associated with listening to unfamiliar accents. Moreover, infants seemingly use surface-level variability in speakers' voices to access the underlying invariant structure. Differences in the way individuals speak can thus serve as a frame of reference to help infants accommodate variation. This makes children's early spoken language processing extremely sophisticated in nature.

Appendix: Infant behavioral techniques

Procedures employed for language and sound discrimination

Conditioned Head Turn Procedure. The goal of this procedure is to train infants to make a head turn each time they detect a sound change. This is implemented by presenting children with a repeating set of sounds (e.g., the sequence /ba ba ba…/), regardless of the infant's response. When a linguistically relevant change occurs (for example, the presentation of /pa/ instead of /ba/), a head turn towards a toy on the child's side is rewarded by the toy lighting up. To help children along in this task, the sound change can, at first, be accompanied by an increase in volume. Over time, this cue fades out, such that the only information signaling the change is the phonetic difference between the speech sounds.

Habituation Procedure. In this procedure, sound presentation is dependent on infants' behavioral responses (looking at the source of the sound, or sucking on a special pacifier). There are two phases to habituation studies. During the initial habituation phase, infants are presented with one or multiple stimuli drawn from a category (e.g. different tokens of the same vowel, or different sentences spoken in the same language) until their interest (measured as their looks at the source of the sound or number of sucks on the pacifier) declines. This is taken to indicate that they have encoded the key features common to the stimuli (e.g., the phonological structure of the language present in the sentences), and are ready to process new information. In a second phase, infants are presented with tokens that belong to a new category. If they increase their attention, and hence dishabituate, this indicates that they have noticed the difference between the two types of presented tokens and can distinguish them. Studies using looking time often have a within-participant design, measuring both responses to new tokens of the habituated category and responses to tokens of a new category in different test trials. They sometimes also contain visual information of the speaker pronouncing the stimuli. Studies relying on sucking responses, by contrast, tend to use between-subject comparisons, whereby one group of infants experiences no change and thus acts as control. They do not have a visual component. Typically, infants in all of these implementations show a novelty preference, reacting more strongly to tokens of a new vowel or passages in a new linguistic variety than to tokens of the habituated category.

Preference Procedure. In this procedure, sound presentation is dependent on infants' behavioral responses (looking at the source of the sound or sucking on a special pacifier). There is typically only one phase to preference studies, although the test phase can be preceded by a familiarization phase. During the test phase, infants are presented with alternating trials that each contain tokens of one type; for example, sentences in the infant's native language versus an unfamiliar language. Sometimes, visual information is available as well. A significant difference in listening times (measured as infants' looks at the source of the sound or number of sucks on the pacifier), revealing a preference for one variant over the other, is interpreted as a sign of discrimination.

Procedures employed for word form recognition and learning

Frequent Word Form Procedure. Infants tested in this paradigm are presented with two types of trials: In familiar word trials, children hear lists of words that occur frequently in speech directed to infants (e.g., *ball, diaper*). By contrast, in unfamiliar word trials, lists of phonotactically legal nonwords (e.g., *dimma*) or real, but rarely occurring words in infant-directed speech (e.g., *feline*) are presented. Alternatively, the list of unfamiliar words may consist of mispronunciations of the likely-known words. Frequently occurring implementations involve either a central fixation screen or a head turn preference set-up with lights positioned in front of the infant as well as on each of the infant's sides. In all cases, children's attention to these items is assessed through orientation times towards the sound source (coming from the direction of the screen or from a blinking side light).

Word Segmentation Procedure. This procedure consists of two phases both of which only present sounds when the infant attends to the source of the sound (a flashing light in case of a head turn implementation or an abstract image shown on a central screen). In the familiarization phase, infants typically hear two repeating word forms presented in isolation. Once children have accumulated a preset amount of listening time, they proceed to the subsequent test phase. In this test phase, children are presented with passages that either do or do not contain the familiarized word forms. Sometimes, the order of words in isolation and words in passages is switched, such that children are familiarized with word forms presented in sentence context and tested on familiarized and unfamiliarized word forms in isolation.

Procedures employed for word-to-image mapping and learning

Intermodal Preferential Looking Procedure (IPLP). This procedure tests word recognition. In the typical IPLP, two images are shown side-by-side on a screen in front of the child. During the presentation of the images, one of them – the target – is named. In many studies, the words and their referents are selected to have a high probability of being well-known to the children tested in the procedure. Sometimes, words are purposely mispronounced. This allows researchers to study the phonetic specification in children's lexical representations. Competitor images may also be well-known words, but sometimes unknown objects (e.g., a rare tool) are used. A greater proportion of looks towards the labeled picture as opposed to the unlabeled one is taken as evidence for the child knowing the word. By examining the time course of children's looking patterns, researchers can furthermore compare the efficiency of word recognition in different conditions. In word learning tasks, the word recognition phase is preceded by a training phase where infants are taught a novel word. This teaching phase can take the form of labeling trials, where the word form is played and a single (or at least unambiguous) image is shown on the screen or it can be conducted in-person.

Switch task. In this procedure, children are first presented with two types of alternating trials. In each trial, the image of a given object displayed on a central screen is paired with a label (e.g., one novel object is paired with *lif* and the other with *neem*). Children are presented with these two word-object pairs until their interest, measured by their looks toward the screen, drops significantly (i.e. they *habituate*). In the subsequent test phase, a single object is projected on the screen, either accompanied by the same label as before (e.g., *lif* with the *lif*-object) or by the other label (e.g., *neem* with the *lif*-object). If children have successfully encoded the two words, looking times should be longer (children should be surprised) when the label and object mismatch compared to when they are matched. Versions where just a single word-object pair is used are possible as well.

References

Barker, B., and Newman, R. (2004). Listen to your mother! The role of talker familiarity in infant streaming. *Cognition*, *94*(2), B45–B53.

Bergelson, E., and Swingley, D. (2012). At 6–9 months, human infants know the meanings of many common nouns. *Proceedings of the National Academy of Sciences*, *109*(9), 3253–3258.

Bergelson, E., and Swingley, D. (2013). Infant word comprehension: Robust to speaker differences but sensitive to single phoneme changes. Talk presented at the Workshop on Infant Language Development, Donostia – San Sebastian, Spain.

Bergmann, C., ten Bosch, L., Fikkert, P., and Boves, L. (2015). Modelling the noise-robustness of infants' word representations: The impact of previous experience. PLoS ONE 10(7): e0132245.

Best, C. T., and McRoberts, G. W. (2003). Infant perception of non-native consonant contrasts that adults assimilate in different ways. *Language and Speech*, *46*(2-3), 183–216.

Best, C. T., McRoberts, G. W., and Sithole, N. M. (1988). Examination of perceptual reorganization for nonnative speech contrasts: Zulu click discrimination by English-speaking adults and infants. *Journal of Experimental Psychology: Human Perception and Performance, 14*(3), 345–360.

Best, C. T., Tyler, M. D., Gooding, T. N., Orlando, C. B., and Quann, C. A. (2009). Development of phonological constancy: Toddlers' perception of native- and Jamaican-accented words. *Psychological Science*, *20*(5), 539–542.

Bortfeld, H., Morgan, J. L., Golinkoff, R. M., and Rathbun, K. (2005). Mommy and me: Familiar names help launch babies into speech-stream segmentation. *Psychological Science*, *16*(4), 298–304.

Bosch, L., Figueras, M., Teixidó, M., and Ramon-Casas, M. (2013). Rapid gains in segmenting fluent speech when words match the rhythmic unit: evidence from infants acquiring syllable-timed languages. *Frontiers in Psychology*, *4*, 106.

Bradlow, A. R., and Bent, T. (2008). Perceptual adaptation to non-native speech. *Cognition*, *106*(2), 707–729.

Butler, J., Floccia, C., Goslin, J., and Panneton, R. (2011). Infants' discrimination of familiar and unfamiliar accents in speech. *Infancy*, *16*(4), 392–417.

Clarke, C. M., and Garrett, M. F. (2004). Rapid adaptation to foreign-accented English. *The Journal of the Acoustical Society of America*, *116*(6), 3647–3658.

Cristia, A., Seidl, A., Vaughn, C., Schmale, R., Bradlow, A., and Floccia, C. (2012). Linguistic processing of accented speech across the lifespan. *Frontiers in Cognition*, *3*, 479.

Curtin, S. A., Fennell, C., and Escudero, P. (2009). Weighting of vowel cues explains patterns of word-object associative learning. *Developmental Science*, *12*(5), 725– 731.

Dahan, D., Drucker, S. J., and Scarborough, R. A. (2008). Talker adaptation in speech perception: Adjusting the signal or the representations? *Cognition*, *108*(3), 710–718.

DeCasper, A. J., and Fifer, W. P. (1980). Of human bonding: Newborns prefer their mothers' voices. *Science*, *208*(4448), 1174–1176.

Dehaene-Lambertz, G., Montavont, A., Jobert, A., Allirol, L., Dubois, J., Hertz-Pannier, L., and Dehaene, S. (2010). Language or music, mother or Mozart? Structural and environmental influences on infants' language networks. *Brain and Language*, *114*(2), 53–65.

Durrant, S., Delle Luche, C., Cattani, A., and Floccia, C. (2015). Monodialectal and multidialectal infants' representation of familiar words. *Journal of Child Language, 42(2)*, 447–465.

Escudero, P., Best, C. T., Kitamura, C., and Mulak, K. E. (2014). Magnitude of phonetic distinction predicts success at early word learning in native and non-native accents. *Frontiers in Psychology*, *5*, 1059.

Fennell, C. T., and Waxman, S. R. (2010). What paradox? Referential cues allow for infant use of phonetic detail in word learning. *Child Development*, *81*(5), 1376–1383.

Fennell, C. T., and Werker, J. F. (2003). Early word learners' ability to access phonetic detail in well-known words. *Language and Speech*, *46*(2-3), 245–264.

Fernald, A., Pinto, J. P., Swingley, D., Weinbergy, A., and McRoberts, G. W. (1998). Rapid gains in speed of verbal processing by infants in the 2nd year. *Psychological Science*, *9*(3), 228–231.

Floccia, C., Delle Luche, C., Durrant, S., Butler, J., and Goslin, J. (2012). Parent or community: Where do 20-month-olds exposed to two accents acquire their representation of words? *Cognition, 124*(1), 95–100.

Floccia, C., Goslin, J., Girard, F., and Konopczynski, G. (2006). Does a regional accent perturb speech processing? *Journal of Experimental Psychology: Human Perception and Performance, 32*(5), 1276–1293.

Galle, M. E., Apfelbaum, K. S., and McMurray, B. (2015). The role of single talker acoustic variation in early word learning. *Language Learning and Development, 11*(1), 66–79.

Goldinger, S. D. (1996). Words and voices: episodic traces in spoken word identification and recognition memory. *Journal of Experimental Psychology: Learning, Memory, and Cognition*, 22(5), 1166–1183.

Goldinger, S. D. (1998). Echoes of echoes? An episodic theory of lexical access. *Psychological Review, 105*(2), 251–279.

Goldinger, S. D. (2007). A complementary-systems approach to abstract and episodic speech perception. In J. Trouvain and W. J. Barry (Eds.), *Proceedings of the 16th International Congress of Phonetic Sciences* (pp. 49–54). Dudweiler, Germany: Pirrot.

Hallé, P. A., and De Boysson-Bardies, B. (1994). Emergence of an early receptive lexicon: Infants' recognition of words. *Infant Behavior and Development*, 17(2), 119–129.

Havy, M., and Nazzi, T. (2009). Better processing of consonantal over vocalic information in word learning at 16 months of age. *Infancy, 14*(4), 439–456.

Hepper, P. G., Scott, D., and Shahidullah, S. (1993). Newborn and fetal response to maternal voice. *Journal of Reproductive and Infant Psychology, 11*(3), 147–153.

Houston, D. M. (1999). *The role of talker variability in infant word representations* (Unpublished doctoral dissertation). The Johns Hopkins University, Baltimore, MD.

Houston, D. M., and Jusczyk, P. W. (2000). The role of talker-specific information in word segmentation by infants. *Journal of Experimental Psychology: Human Perception and Performance*, 26(5), 1570–1582.

Johnson, E. K., Seidl, A., and Tyler, M. D. (2014). The edge factor in early word segmentation: utterance-level prosody enables word form extraction by 6-month-olds. *PloS One*, 9(1), e83546.

Jusczyk, P. W. (1997). *The discovery of spoken language*. Cambridge, MA: MIT press.

Jusczyk, P. W., and Aslin, R. N. (1995). Infants' detection of the sound patterns of words in fluent speech. *Cognitive Psychology*, *29*(1), 1–23.

Jusczyk, P. W., and Hohne, E. A. (1997). Infants' memory for spoken words. *Science*, *277*(5334), 1984–1986.

Jusczyk, P. W., Houston, D. M., and Newsome, M. (1999). The beginnings of word segmentation in English-learning infants. *Cognitive Psychology*, *39*(3), 159–207.

Jusczyk, P. W., Pisoni, D. B., and Mullennix, J. (1992). Some consequences of stimulus variability on speech processing by 2-month-old infants. *Cognition*, *43*(3), 253–291.

Kinzler, K. D., Corriveau, K. H., and Harris, P. L. (2011). Children's selective trust in native-accented speakers. *Developmental Science*, *14*(1), 106–111.

Kinzler, K. D., Dupoux, E., and Spelke, E. S. (2007). The native language of social cognition. *Proceedings of the National Academy of Sciences*, *104*(30), 12577–12580.

Kinzler, K. D., Shutts, K., DeJesus, J., and Spelke, E. S. (2009). Accent trumps race in guiding children's social preferences. *Social Cognition*, *27*(4), 623–634.

Kitamura, C., Panneton, R., and Best, C. T. (2013). The development of language constancy: Attention to native versus nonnative accents. *Child Development*, *84*(5), 1686–1700.

Kreiman, J., and Sidtis, D. (2011). *Foundations of voice studies: An interdisciplinary approach to voice production and perception*. Malden, MA: John Wiley & Sons.

Kuhl, P. K. (1979). Speech perception in early infancy: Perceptual constancy for spectrally dissimilar vowel categories. *The Journal of the Acoustical Society of America*, *66*(6), 1668–1679.

Kuhl, P. K. (1983). Perception of auditory equivalence classes for speech in early infancy. *Infant Behavior and Development*, *6*(2), 263–285.

Kuhl, P. K., Stevens, E., Hayashi, A., Deguchi, T., Kiritani, S., and Iverson, P. (2006). Infants show a facilitation effect for native language phonetic perception between 6 and 12 months. *Developmental Science*, *9*(2), F13–F21.

Levi, S. V. (2015). Talker familiarity and spoken word recognition in school-age children. *Journal of Child Language 42(4)*, 843–872.

Luce, P. and McLennan, C. (2005). Spoken word recognition: The challenge of variation. In D. Pisoni, D. and R. Remez (Eds.), *The Handbook of Speech Perception* (pp. 591–609). Malden, MA: Blackwell.

Mandel, D. R., Jusczyk, P. W., and Pisoni, D. B. (1995). Infants' recognition of the sound patterns of their own names. *Psychological Science*, 6(5), 314–317.

Mani, N., and Plunkett, K. (2007). Phonological specificity of vowels and consonants in early lexical representations. *Journal of Memory and Language*, *57*(2), 252–272.

Mani, N., and Plunkett, K. (2008). Fourteen-month-olds pay attention to vowels in novel words. *Developmental Science*, *11*(1), 53–59.

Mani, N., and Plunkett, K. (2011). Does size matter? Subsegmental cues to vowel mispronunciation detection. *Journal of Child Language*, *38*(3), 606–627.

Maye, J., Aslin, R., and Tanenhaus, M. (2008). The weckud wetch of the wast: Lexical adaptation to a novel accent. *Cognitive Science*, *32*(3), 543–562.

McMurray, B., and Aslin, R. N. (2005). Infants are sensitive to within-category variation in speech perception. *Cognition*, *95*(2), B15–26.

Mehler, J., Bertoncini, J., Barrière, M., and Jassik-Gerschenfeld, D. (1978). Infant recognition of mother's voice. *Perception*, *7*(5), 491–497.

Mehler, J., Jusczyk, P., Lambertz, G., Halsted, N., Bertoncini, J., and Amiel-Tison, C. (1988). A precursor of language acquisition in young infants. *Cognition*, *29*(2), 143–178.

Moon, C., Cooper, R. P., and Fifer, W. P. (1993). Two-day-olds prefer their native language. *Infant Behavior and Development*, *16*(4), 495–500.

Mulak, K. E., and Best, C. T. (2013). Development of word recognition across speakers and accents. In L. J. Gogate and G. Hollich (Eds.), *Theoretical and computational models of word learning: Trends in psychology and artificial intelligence* (pp. 242–269). Hershey: IGI Global: Robotics Division.

Mulak, K. E., Best, C. T., Tyler, M. D., Kitamura, C., and Irwin, J. R. (2013). Development of phonological constancy: 19-month-olds, but not 15-month-olds, identify words in a non-native regional accent. *Child Development*, *84*(6), 2064–2078.

Naoi, N., Minagawa-Kawai, Y., Kobayashi, A., Takeuchi, K., Nakamura, K., Yamamoto, J., and Kojima, S. (2012). Cerebral responses to infant-directed speech and the effect of talker familiarity. *NeuroImage*, *59*(2), 1735–1744.

Nazzi, T. (2005). Use of phonetic specificity during the acquisition of new words: Differences between consonants and vowels. *Cognition, 98*(1), 13–30.

Nazzi, T., Bertoncini, J., and Mehler, J. (1998). Language discrimination by newborns: toward an understanding of the role of rhythm. *Journal of Experimental Psychology: Human Perception and Performance*, *24*(3), 756

Nazzi, T., Jusczyk, P. W., and Johnson, E. K. (2000). Language discrimination by English-learning 5-month-olds: Effects of rhythm and familiarity. *Journal of Memory and Language*, *43*(1), 1–19.

Newman, R. S. and Jusczyk, P. W. (1996). The cocktail party effect in infants. *Perception and Psychophysics, 58*(8), 1145–1156.

Parise, E., and Csibra, G. (2012). Electrophysiological evidence for the understanding of maternal speech by 9-month-old infants. *Psychological Science*, 728–733.

Pierrehumbert, J. (2006). The next toolkit. *Journal of Phonetics, 34*(6), 516–530.

Purhonen, M., Kilpeläinen-Lees, R., Valkonen-Korhonen, M., Karhu, J., and Lehtonen, J. (2004). Cerebral processing of mother's voice compared to unfamiliar voice in 4-month-old infants. *International Journal of Psychophysiology*, *52*(3), 257–266.

Ramus, F., Nespor, M., and Mehler, J. (1999). Correlates of linguistic rhythm in the speech signal. *Cognition*, *73*(3), 265–292.

Rost, G. C., and McMurray, B. (2009). Speaker variability augments phonological processing in early word learning. *Developmental Science*, *12*(2), 339–349.

Rost, G. C., and McMurray, B. (2010). Finding the signal by adding noise: The role of noncontrastive phonetic variability in early word learning. *Infancy*, *15*(6), 608–635.

Schmale, R., Cristia, A., and Seidl, A. (2012). Toddlers recognize words in an unfamiliar accent after brief exposure. *Developmental Science*, *15*(6), 732–738.

Schmale, R., Cristià, A., Seidl, A., and Johnson, E. K. (2010). Developmental changes in infants' ability to cope with dialect variation in word recognition. *Infancy*, *15*(6), 650–662.

Schmale, R., Hollich, G., and Seidl, A. (2011). Contending with foreign accent in early word learning. *Journal of Child Language*, *38*(5), 1096–1108.

Schmale, R., and Seidl, A. (2009). Accommodating variability in voice and foreign accent: flexibility of early word representations. *Developmental Science*, *12*(4), 583–601.

Schmale, R., Seidl, A. and Cristia, A., (2015). Mechanisms underlying accent accommodation in early word learning: Evidence for general expansion. *Developmental Science 18(4)*, 664–670.

Seidl, A., Onishi, K. H., and Cristia, A. (2014). Talker variation aids young infants' phonotactic learning. *Language Learning and Development*, *10*(4), 297–307.

Singh, L. (2008). Influences of high and low variability on infant word recognition. *Cognition*, *106*(2), 833–870.

Singh, L., Morgan, J. L., and White, K. S. (2004). Preference and processing: The role of speech affect in early spoken word recognition. *Journal of Memory and Language*, *51*(2), 173–189.

Stager, C. L., and Werker, J. F. (1997). Infants listen for more phonetic detail in speech perception than in word-learning tasks. *Nature*, *388*(6640), 381–382.

Swingley, D. (2005). 11-month-olds' knowledge of how familiar words sound. *Developmental Science*, *8*(5), 432–443.

Swingley, D. (2009). Onsets and codas in 1.5-year-olds' word recognition. *Journal of Memory and Language*, *60*(2), 252–269.

Swingley, D., and Aslin, R. N. (2002). Lexical neighborhoods and the word-form representations of 14-month-olds. *Psychological Science*, *13*(5), 480–484.

Tincoff, R., and Jusczyk, P. W. (1999). Some beginnings of word comprehension in 6-month-olds. *Psychological Science*, *10*(2), 172–175.

Tincoff, R., and Jusczyk, P. W. (2012). Six-month-olds comprehend words that refer to parts of the body. *Infancy*, *17*(4), 432–444.

Trude, A. M., and Brown-Schmidt, S. (2012). Talker-specific perceptual adaptation during online speech perception. *Language and Cognitive Processes*, *27*(7-8), 979–1001.

Van der Feest, S. V. H., and Johnson, E. K. (in press). Input driven differences in toddler's peception of a disappearing phonological contrast. *Language Acquisition.*

Van Heugten, M., and Johnson, E. K. (2012). Infants exposed to fluent natural speech succeed at cross-gender word recognition. *Journal of Speech, Language and Hearing Research*, *55*(2), 554–560.

Van Heugten, M., and Johnson, E. K. (2014). Learning to contend with accents in infancy: Benefits of brief speaker exposure. *Journal of Experimental Psychology: General*, *143*(1), 340–350.

Van Heugten, M., Krieger, D. R., and Johnson, E. K. (2015). The developmental trajectory of toddlers' comprehension of unfamiliar regional accents. *Language Learning and Development, 11*(1), 41–65.

Vihman, M. M., Nakai, S., DePaolis, R. A., and Hallé, P. (2004). The role of accentual pattern in early lexical representation. *Journal of Memory and Language*, *50*(3), 336–353.

Werker, J. F., and Curtin, S. (2005). PRIMIR: A developmental framework of infant speech processing. *Language Learning and Development*, *1*(2), 197–234.

Werker, J. F., Fennell, C. T., Corcoran, K. M., and Stager, C. L. (2002). Infants' ability to learn phonetically similar words: Effects of age and vocabulary size. *Infancy*, *3*(1), 1–30.

Werker, J. F., and Tees, R. C. (1984). Cross-language speech perception: Evidence for perceptual reorganization during the first year of life. *Infant Behavior and Development*, *7*(1), 49–63.

White, K. S., and Aslin, R. N. (2011). Adaptation to novel accents by toddlers. *Developmental Science*, *14*(2), 372–384.

White, K. S., and Morgan, J. L. (2008). Sub-segmental detail in early lexical representations. *Journal of Memory and Language*, *59*(1), 114–132.

Yoshida, K. A., Fennell, C. T., Swingley, D., and Werker, J. F. (2009). Fourteen-month-old infants learn similar-sounding words. *Developmental Science*, *12*(3), 412–418.

Benjamin Swets

Grand Valley State University

Psycholinguistics and Planning: A Focus on Individual Differences

Abstract: Researchers in the field of psycholinguistics, and especially language production, tend to use experimental research methods to test theories of models of processing. In doing so, we sometimes overlook systematic variance in task performance that is due to individual differences. One area that could benefit from more work on individual differences is in research concerning the mental mechanisms governing the scope of advance speech planning. In this chapter, I will summarize some of the research I have conducted with colleagues that has explored the utility of the individual differences approach. First, I will show how individual differences approaches can capture a good deal of variance that other more traditional variables might miss. I will then offer some data consistent with the idea that the scope of planning in language production varies not just across experimentally manipulated conditions, but also among individuals. Following this I will argue that this individual differences approach allows for some theoretical advances regarding the general role of working memory in language processing. I will conclude by outlining additional opportunities to conduct individual differences research in language production, with some notes to take appropriate caution doing so.

1. Psychology and variance

The overall purpose of cognitive psychological research is to discover *systematic variance* in behaviors that can help us to infer the nature of our mental processes and representations. In psycholinguistics, researchers have typically proceeded by searching for systematic variance across situations by manipulating independent variables experimentally. The virtues of this approach are clear, in that they provide the means to assess the causal relationships between variables that allow researchers to arrive at meaningful explanations (models and theories) of psycholinguistic phenomena. The focus of this chapter is on systematic variance in language processing *among individuals (individual differences)*.

The well known limitation of the individual differences approach is that it is inherently correlational. By merely associating variables rather than manipulating them, researchers fall short of causal explanations. But this type of research also does something very well that the typical psycholinguistic experiment cannot: It can systematically account for variance that occurs among individuals, which generally is error variance (variance that cannot be accounted for) in most experiments. To illustrate the utility of the individual differences approach in psycholinguistics, I presently review a case in which individual differences approaches yielded insights that experimental approaches alone could not have provided. We will begin in the domain of sentence comprehension and then turn to sentence production.

2. An illustrative example

One theoretical focus of psycholinguistic research on sentence comprehension has been on the manner in which a parser decides what to do with new, ambiguous constituents. Frazier (1987), as part of Garden Path Theory, a modular, syntax-first account of parsing, postulated Late Closure, a universal parsing principle: "If grammatically possible, attach new items into the clause or phrase currently being processed." Take sentence (1) below. One possible parse of this sentence is to associate the relative clause with the first noun phrase (NP1), "the sister". This "high" attachment interpretation implies that it was the sister who shot herself on the balcony. According to research by Frazier (1979), English speakers instead prefer to associate the relative clause with the second noun phrase (NP2). According to this "low" type of attachment, it was the actress who shot herself on the balcony. According to Frazier's Garden Path model, this preference exists because the Late Closure heuristic makes a decision based on syntax alone that the relative clause must be part of the currently "open" phrase, which in the case of (1) is "the actress". Importantly, Garden Path model posited that such parsing strategies should be universal, holding that all languages ought to show the same preference in similar constructions.

(1) The sister of the actress who shot herself on the balcony was under investigation.

Cuetos and Mitchell (1988) showed that other languages, such as Spanish and Dutch, showed an NP1 preference (Cuetos and Mitchell, 1988; Brybaert and Mitchell, 1996). Because preferences varied across languages, this line of research undermined that assumption of Late Closure as a universal parsing strategy. Without such universality, the viability of the syntax-first Garden Path model was reduced.

Subsequent research has demonstrated that despite the theoretical importance of these cross-linguistic differences, there is even more variability in attachment preferences among individual speakers of the same language (Swets et al., 2007). In the research that showed these results, we conducted two studies. In the first, we administered a reading span task to measure working memory and an offline relative clause attachment task to each participant from English speaking (n = 150) and Dutch speaking (n = 96) populations. The tasks were administered to large samples of subjects because the statistical analyses we were conducting, including factor analysis and structural equation modeling, required large samples to detect theoretically interesting effects. For the relative clause attachment task, participants viewed sentences such as (1) on a screen, and then were asked forced choice questions about them that indicated a NP1 or NP2 attachment decision.

2) Who was shot on the balcony? (the sister / the actress)

The reading span task we administered was a modified version of the Daneman and Carpenter (1980) task. In the task, participants tried to remember lists of 3 to 6 words as they judged whether a series of sentences made sense. Underneath each sentence that they read on the screen in front of them, a word appeared in red. Participants were to circle *YES* on an answer sheet if the sentence made sense, and *NO* if it did not, and remember the word in red for later. After 3 to 6 sentence judgments, three question marks appeared on the screen, and participants were to turn the page on the answer sheet and write each of the red words down in the order in which they appeared.

We found that reading span predicted attachment preferences, although the direction of the relationship was surprising. Participants with lower working memory scores (low-spans) tended to attach relative clauses high, to NP1, and participants with higher working memory scores (high-spans) tended to attach low, to NP2. This trend held for both English and Dutch speakers, even though the overall attachment preferences differed between languages. In other words, consistent with previous research, English speakers still preferred NP2 attachment overall, and Dutch speakers still preferred NP1 attachment overall, but within each language, there were rather large systematic individual differences that exceeded those cross-linguistic differences. When we computed the effect size of language spoken on attachment preferences, we found that Cohen's $d = .29$, which is regarded as a "small" effect. The effect size of the individual differences in attachment preference, on the other hand, was "large", Cohen's $d = .72$ in the English sample and .90 in the Dutch sample. To interpret these statistics a bit more, this means being a speaker of Dutch versus English accounts for about 30% of a standard deviation of the measure. But individual-specific verbal working memory score accounts for between 70–90% of a standard deviation. In short, individual differences in attachment preferences account for 3 times as much variance as cross-linguistic differences. One overall implication of this finding is that psycholinguistic processing principles once thought to be inflexible and automatic, such as Late Closure, can be shown to be highly flexible when examining individual differences. The second study in this line of research sought an explanation of the effects showing a NP2 preference for high-span participants. I will return to this study later to report those findings and their implications, but for now, I turn to another process once thought to be highly rigid and inflexible: sentence planning scope.

3. Variation in the scope of sentence planning

3.1. Inflexible units in sentence planning: A critical review

The most comprehensive examination of the functional language production system is Levelt's model (Levelt, 1989; Bock and Levelt, 1994). It assumes a language production system with insular, sequential levels of processing: information at one level cannot be computed until receiving as input the output from the preceding level. First is the "message" level, the stage at which the basic semantic proposition the speaker intends to utter is composed. The next stage in processing is "grammatical encoding", the accessing of non-phonological word information (meaning and syntactic category) plus the structuring of these so-called "lemmas" into their phrasal positions to produce surface structure. This surface structure representation is passed to "phonological encoding", which retrieves word forms and creates a prosodic structure. Articulation is the final stage. A hallmark of this model is that each processing level works in parallel with the other levels in a pipe line mode. After the syntactic level outputs to the phonological level, syntax works on another piece of semantic input as phonology deals with the initial syntactic piece. Sentence production is termed incremental because of this parallelism and because as linguistic representations are shunted from one processing level to another, the range of operation is not over an entire sentence. Rather, speakers plan in increments, packaging small pieces of information together before that chunk is sent to the following level.

One aspect of this model that for a long time had been relatively uncontroversial is the claim that planning at many of these levels of representation is automatic: That is, such planning doesn't require any processing resources such as working memory (Levelt, 1989; reviewed in Garrod and Pickering, 2007). Indeed, there would be obvious advantages found in a system that does not need cognitive resources to operate. However, in order to achieve that kind of freedom from resources, the planning system would have to sacrifice something: flexibility in the extent to which utterances can be planned in advance. Hence, such models assume that planning scope is stable, or inflexible, and architecturally minimal.

This assumption of automatic, inflexible units of planning has been accompanied by an empirical search for what those units might be. For example, Smith and Wheeldon (2001) conducted a set of experiments to test whether there is costly syntactic planning before speech onset. In their experiments, sentences like "The spoon and the car move up" were used to prime the production of syntactically related sentences (Experiments 1–5) in picture description tasks. When participants uttered sentences that were syntactically like the previous sentence, a reliable 50 ms advantage to begin speaking was found. The measure indicated how much time had been saved in the planning of the syntactic frame of the sentence. Smith and Wheeldon also tested the scope of this effect and found that it only held for the first phrase of an utterance, leading them to conclude that phrases are the automatically planned units of grammatical encoding. Another study (Griffin, 2001) showed that when speakers described scenes with 3 objects (at positions A, B and C) using sentences like "The A and the B are above the C", only the word frequency of the object at position A influenced speech latency. Griffin concluded that speech planning is automatic, with a minimal, phrasal planning scope.

The notion that it will be possible to find fixed, automatic planning units is likely flawed for several reasons. As Levelt (1989) points out, in the time between 1967 and 1989, at least 18 different speech "planning units" (also sometimes referred to as "lookahead" or "scope") of varying sizes had been proposed in the literature, leading Levelt to remark "...there is no single unit of talk" (Levelt, 1989, p. 23). Although this quote applies broadly across the different levels of representation listed above, there has since also been disagreement regarding what length those units might be, even within the same level of representation. For example, within the domain of grammatical encoding, although there are several studies that support the architecturally sub-clausal or phrasal view of incremental planning (Schriefers et al., 1998; Smith and Wheeldon, 1999, 2001), other studies show that the scope of planning extends to as much as a whole clause (Christianson and Ferreira, 2005; Ford and Holmes, 1978). Second, the studies that reveal minimal planning scopes neglect to apply pressures on planning processes to see whether the scope of planning can be flexibly pushed around in different situations. Lastly, there has been very little research into individual differences in planning scope. These reservations have given rise to an alternative account to rigid incremental planning during speech production: the flexible incrementality view. Much recent research has demonstrated that although there are circumstances when speech planning proceeds very incrementally—that is, bit by bit—there are also circumstances that dictate more planning to be done in advance (Costa and Caramazza; 2002; Damian and Dumay, 2007; Ferreira and Swets, 2002, 2005; Fuchs et al., 2013; Konopka and Meyer, 2010; Korvorst et al. 2006; Schriefers and Teruel, 1999a; Wagner et al., 2010). I will presently review some of these circumstances, including manipulations of time pressure and cognitive load.

3.2. Variation across situations

One way my colleagues and I have demonstrated this flexibility in planning scope is by having speakers produce sentences that can have some aspect of complexity manipulated very late in the sentence (e.g., Ferreira and Swets, 2002, 2005). In one such study (Ferreira and Swets, 2002), participants produced sentences as they solved simple math problems of subtly differing complexity. In the so-called "easy" condition, the speaker would see "21 + 22" on the screen, and the target utterance was "The answer is 43." In the "hard" condition, the speaker would see "25 + 23" on the screen, and the target utterance was "The answer is 48." Note that it takes people reliably longer to calculate the hard problem than the easy problem—sums totalling between 6 and 9 (such as the *8* in *48*) take reliably longer to compute than sums totaling 5 or less (such as the *3* in *43*) (Ashcraft, 1992). By exploiting this tendency, we estimated how far speakers planned their sentences in advance by measuring when speech slowed down in the hard condition relative to the easy condition. In Experiment 1, although we asked speakers to give their answers as quickly and as accurately as possible, the speakers were free to begin speaking whenever they chose. To assess the location at which speakers slowed down to plan for the difficult problems relative to the easy problems, we measured initiation time to begin speaking as well as the durations of the subsequent sections of the utterances, including each word of "The answer is" and both the 10s and 1s place of the arithmetic answer. Under these conditions, speakers slowed in the hard condition relative to the easy condition only in their latency to begin speaking. Once articulation began, we found no effects of problem difficulty on speech duration, implying that speakers had planned the entire sentence, including the solution to the addition problem, prior to the onset of speech.

In Experiment 2, we introduced an explicit deadline to begin speaking. During and after the practice phase of this experiment, a timer began counting down as soon as the arithmetic problem appeared on the screen. If participants did not respond before the timer finished, they heard a "beep" sound that indicated the deadline to begin speaking had passed. Under this deadline, effects of problem difficulty were found at each hand-measured section of an utterance, including the initiation time, "The answer is", and the answer itself. In other words, even though speakers were doing some long-distance planning before beginning to speak, they were leaving some of the planning of these sentences for later. We concluded that the situation of time pressure had some influence on the scope of planning.

Since this research, several other studies (Damian and Dumay, 2007; Ferreira and Swets, 2005; Fuchs et al., 2013; Konopka and Meyer, 2010; Korvorst et al. 2006; Wagner et al., 2010) have demonstrated how planning scope can vary across situations. One noteworthy example of this was a series of experiments reported by Wagner et al. (2010), who found that increased task load reduces the scope of grammatical encoding. In that research, participants were either in the situation of having a low or high task load concurrently with the production task. The situation of increased load reduced planning scope. More recently, Fuchs et al. (2013) reported that different *measures* of planning reveal different simultaneous planning scopes. Whereas initial f0 peak was only sensitive to local planning considerations, measures such as pause duration, inhalation duration, and inhalation depth seemed sensitive to longer-distance planning effects.

3.3. Variation among individuals

This evidence that the scope of planning is flexibly adaptive to situational manipulations may indicate a planning system that is also adaptive to individual differences. Both the Wagner et al. (2010) and Fuchs et al. (2013) studies allude to the possibility that a large amount of variance in planning scope occurs among individuals. Fuchs et al. (2013) for example found large speaker-specific variation in planning as measured by breath inhalation depth prior to articulation. Whereas some of the participants would inhale very deeply prior to a long sentence compared to a short sentence (indicating a long scope of planning), some participants inhaled nearly equally in long and short sentences. Figure 1 illustrates this phenomenon, showing how the difference in inhalation between long and short sentences varies among the individual participants. The resulting figure illustrates a phenomenon that I mentioned at the outset, which is that most experimental studies are unable to account for variance among individuals. The data points in Figure 1 beg to be systematized and ordered on some dimension. As presented, these data points represent error variance: variance among participants that cannot be accounted for. But the goal of psychology, as mentioned earlier, is to find systematic variance—to find some way to straighten a seemingly random array of dots. Here is where correlational, individual differences techniques find their utility.

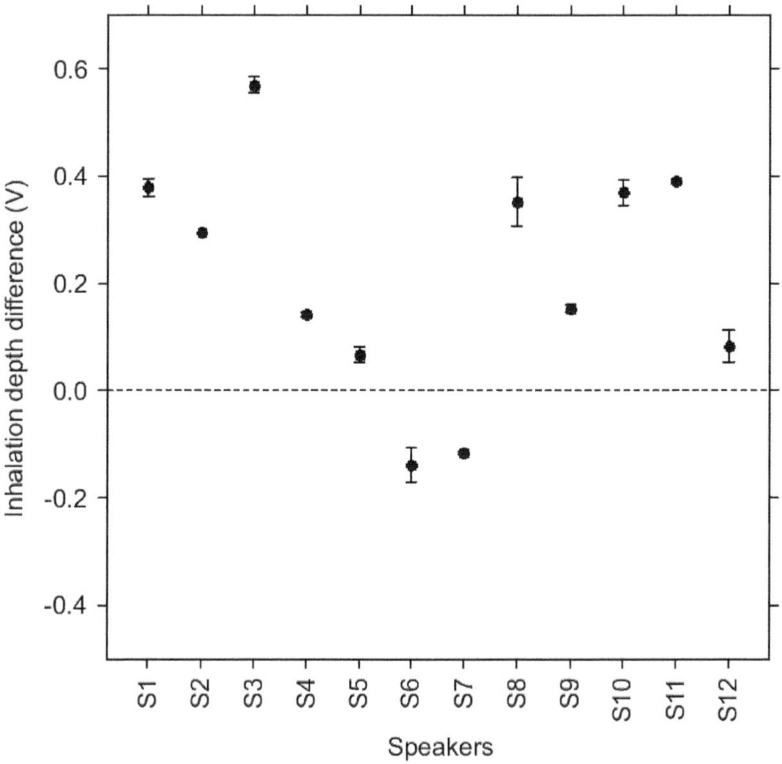

Figure 1. Results presented in Fuchs et al. (2013) illustrating the difference in inhalation depth between long and short sentences. Each point represents an individual participant. Reprinted from Journal of Phonetics, 41, *Fuchs, S., Petrone, C., Krivokapić, J., and Hoole, P., Acoustic and respiratory evidence for utterance planning in German, 29–47, Copyright 2013, with permission from Elsevier.*

One such correlational variable that can help systematically account for some of this variation among individuals, and "straighten out" figures like the one above, is planning time. Wagner et al. (2010), in an analysis done to help rule out alternative explanations of their results, found that the amount of time taken to begin articulating a given sentence reliably predicted the distance of the interference effects that indicated the scope of syntactic planning: The more quickly speakers initiated articulation, the less likely they were to show long-distance interference effects in their planning. More research has recently emerged showing similar patterns: If an individual speaker chooses a strategy of speaking sooner rather than later, there are associated reductions in planning scope (Gillespie and Pearlmutter, 2011; Lange and Laganaro, 2014). In one such experiment, Gillespie and Pearlmutter (2011) elicited sentences that could potentially elicit subject-verb agreement errors such as *The apple near the pies was/*were*, and argued that increased numbers of errors suggests a longer scope of advance planning. Results showed that speakers with higher average speech onset time produced more such errors, indicating a longer scope of planning. Likewise, in an experiment that measured the scope of phonological planning by priming the first and second elements (noun-adjective or adjective-noun) in a sentence-initial noun phrase, Lange and Laganaro (2014) showed that only the participants who delayed the beginnings of their utterances showed priming beyond the first element. Such research suggests that those who take additional time tend to plan more material in advance of speech.

The other primary approach to examining planning scope variation among individuals has been to examine working memory capacity (Petrone et al., 2011; Swets et al., 2014). Although several previous studies employing a variety of approaches have demonstrated that higher-level language production, including grammatical encoding, is supported by working memory resources (Hartsuiker and Barkhuysen, 2006; Horton and Spieler, 2007; Kellogg et al., 2007; Kemper et al., 2003; Kemper and Sumner, 2001; Slevc, 2011), none of these studies had examined individual differences in planning scope. Petrone and colleagues (2011) found that working memory predicted the pitch of speakers' voices to begin articulating phrases of different complexity, suggesting that speakers with more working memory may have a greater planning scope. The rationale of using initial utterance pitch as an indicator of planning scope is that longer phrases are associated with a greater pitch declination from start to finish. Speakers who can plan more in advance are those who are more likely to begin their sentences at a higher pitch to anticipate the upcoming declination. The results showed that high span speakers began articulation of complex subject phrases at a higher pitch than low span speakers. One interesting note about this finding to which I will return later is that despite this observed difference in apparent planning scope based on working memory, preparation time to begin articulation was equivalent among the groups (Petrone et al., 2011).

4. The role of working memory in planning

Speakers often utter sentences in circumstances of referential ambiguity, and in such circumstances, careful sentence planning can spell the difference between effective and ineffective communication. Suppose there is a carpenter's assistant holding two hammers, and a harried carpenter who asks that assistant to "Hand me the hammer." Had the carpenter planned more carefully, a more optimal sentence might be "Hand me the smaller hammer." The hypothesis of a recent study I conducted with colleagues (Swets et al., 2014) was to investigate whether individual differences in working memory predict variation in the scope of advance sentence planning in such circumstances of referential ambiguity. We reasoned that someone with high working memory capacity might be capable of both gathering important information about ambiguities and integrating such information into their speech plans.

A secondary aim of the study was to identify what kind of role working memory plays in the planning process. One view of the role working memory might play in planning scope is that it affords a storage space for the messages one generates while planning. On this view, with limited working memory, only small increments can be planned at one time because a lower capacity prevents the storage of larger plans. According to this hypothesis, speakers with more working memory will plan more content in advance, but like the speakers in Gillespie and Pearlmutter (2011), Lange and Laganaro (2014), and Wagner et al. (2010), they should also have to spend more time creating those larger plans. We also tested an alternative view of the role of working memory that gives it not just storage functions, but also efficiency functions. According to this hypothesis, working memory performs the job not only of simple storage of generated message plans, but also of integrating and packaging linguistic information in a temporally efficient manner. As such, speakers with more working memory should be able to plan more of a sentence in advance, but do so without taking up additional time. Alternatively, they might plan the same amount as low-span speakers, but in less time (see Heitz and Engle, 2007 for a similar effect in the working memory literature). A prior result obtained in our lab had suggested this possibility of increased efficiency (Swets et al., 2013). In that study, we observed that speakers with co-present conversational partners provided more detailed descriptions of ambiguous objects, reflecting more careful planning, than speakers without a co-present addressee. However, the groups did not differ on initiation times. Perhaps participants with additional WM capacity are likewise capable of doing more planning on an equivalent time scale.

In order to test these hypotheses, we used eye tracking to measure the extent to which speakers inspected more advance regions of a visual display before beginning to describe it. The description paradigm we used was inspired by Brown-Schmidt and Tanenhaus (2006), who had previously used images containing two referents so similar to each other that to describe the picture to an interlocutor, one must distinguish between the two referents by using a modifying expression. We also measured individual differences in working memory via a reading span task (using the same materials as Swets et al., 2007). We hypothesized that working memory is used to prepare and store larger utterance plans, suggesting that people with high working memory capacity should literally look further ahead into the picture when planning high-level sentence information than people with low working memory capacity.

Figure 2 presents examples of the two types of displays we showed to participants in this study. In experimental conditions, we showed the cat with four legs in the first position, and the cat with three legs in the third position. Control conditions featured a different object in the third position, such as a wheel.

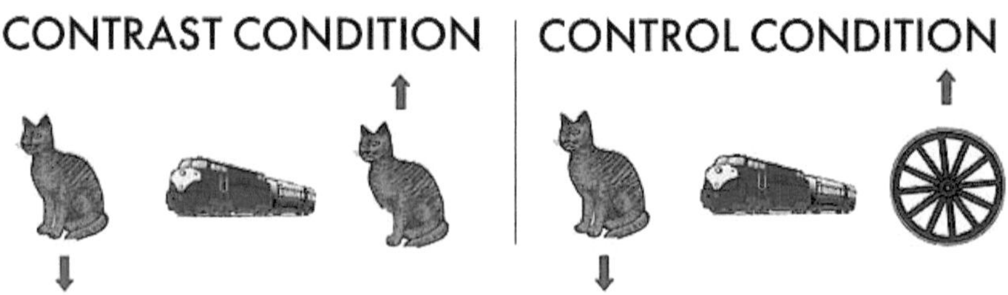

Figure 2. Examples of contrast and control displays in Swets et al. (2014)

The study consisted of two phases. We assessed working memory via reading span in phase I. In phase II, a sample of participants from phase I who demonstrated a wide range of reading span scores returned to act as Directors in a matching game. During the game, Directors produced utterances that mentioned the three objects and the directions of their movement (as indicated by the arrows). For example, in response to the contrast display, a participant might say, "The four-legged cat moves below the train and the three-legged cat moves above the train," though certainly other descriptions that fit the target frame were possible.

Directors understood that the purpose of their utterances was to allow Matchers to manipulate items on a grid displayed on the Matchers' own computer (see Figure 3). Because Matchers had the same cats in their displays as the Director, it was important for the Director to modify both the first noun, that corresponded to the object in Region 1, and the third noun, that corresponded to the object in Region 3 (see Figure 4). For control conditions, there was no need to modify either the first noun phrase (N1), cat, or the third noun phrase (N3), wheel, as the Matcher also saw only one cat in that condition.

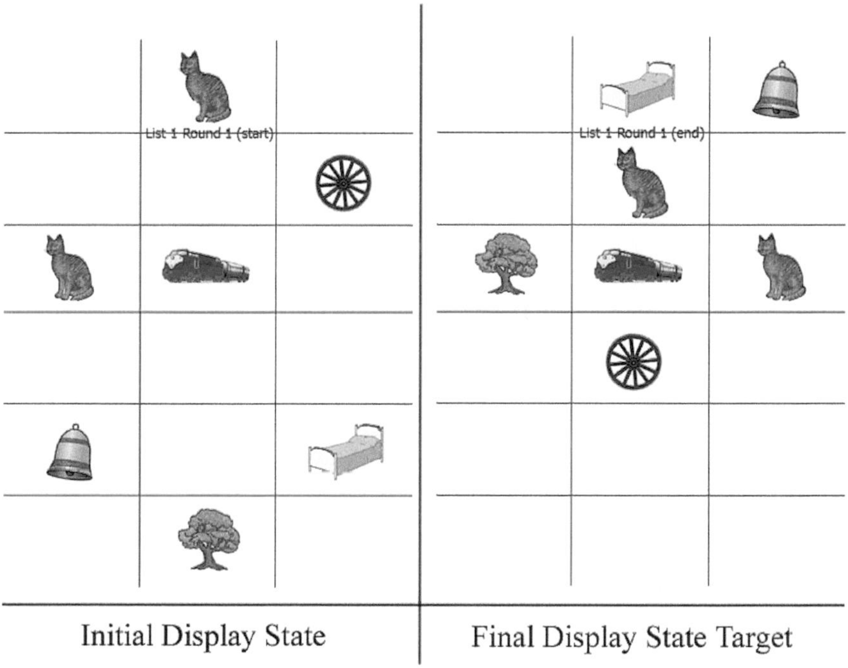

Figure 3. Examples of Matchers' displays from Swets et al. (2014).

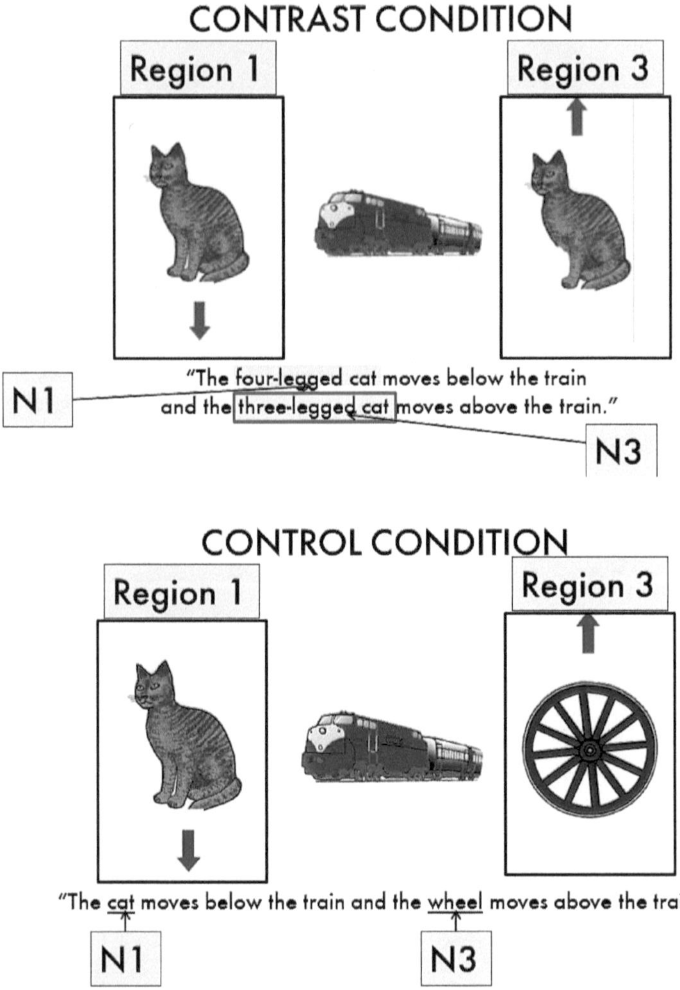

Figure 4. Depictions of visual Regions 1 and 3, and noun phrases 1 and 3 (N1 and N3) for the contrast and control conditions of Swets et al. (2014).

Our dependent measures included initiation time, or the time taken to begin speaking, and fixation patterns, or where people were looking during particular windows of time. We also examined the content of N1 and N3 descriptions by coding whether participants modified N1 and N3. We treated working memory (WM) as a continuous measure to avoid artificial dichotomization. Results were analyzed using linear mixed effects models in R, with WM and display type entered as interactive fixed effects, and participants and items entered as random effects.

According to the general hypothesis that working memory supports advance planning processes, we predicted that we would observe correlations between working memory and our measures of advance planning. Specifically, working memory should correlate with the tendency to look at the contrast object in Region 3 (e.g., the three-legged cat) before speaking and with the tendency to modify N1 (e.g., *The four-legged cat* rather than *The cat*) early on in the sentence. Analysis of the time course in which these additional looks and modifications took place was intended to help distinguish between WM as a simple capacity limitation or a more active, efficient integration process. If high span participants require more time to plan additional content prior to speech, then WM could be viewed simply as storage for larger plans. If high span participants take the same (or less) time to plan more content than low spans, then the role of WM is more complex in that it also invokes temporally efficient packaging.

In an example of a typical observed trial, a Director with high reading span describing the cat display shown in Figure 4 might fixate both cat one and cat two before articulating a description that modifies both the first noun, N1, and the third noun, N3. A low-span individual describing the same display might fail to fixate the contrast cat in Region 3 before beginning to speak, and then fail to subsequently modify N1. You will see from these data that for contrast displays, high spans were more apt than low spans to not only fixate Region 3 before speaking, but also include a modification of N1 that helped listeners immediately distinguish between two possible referents.

The first analysis presented in Figure 5 is the amount of time Directors took to begin speaking. The figure shows that during contrast trials, working memory did not predict the amount of time to begin speaking: Everyone took about 2.5 seconds on average. On the other hand, during control trials, when there was less planning work to do, high spans began speaking more quickly than low spans. Consistent with the efficient capacity account, working memory allowed speakers in these control trials to plan equivalent content in less time.

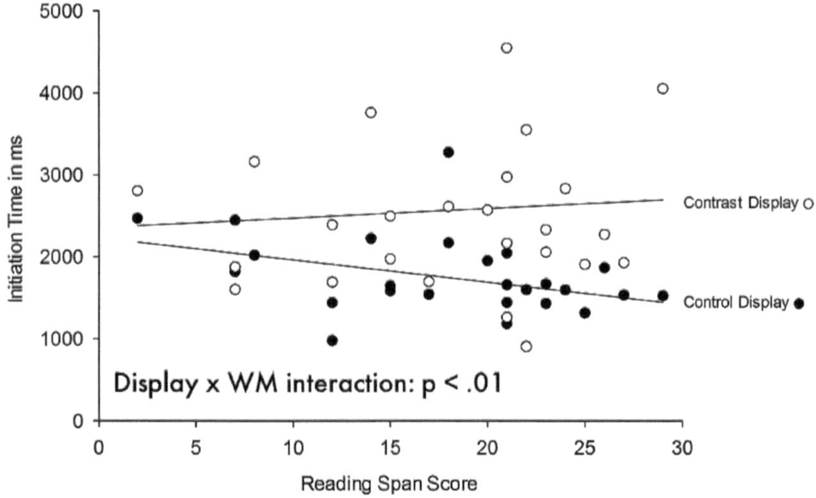

Figure 5. Initiation time results from Swets et al. (2014). Reprinted from Language and Cognition, *6, Issue 01, 2014, 12–44, Swets, B., Jacovina, M.E., and Gerrig, R.J. Individual differences in the scope of speech planning: Evidence from eye movements. Copyright © 2014 UK Cognitive Linguistics Association. Reprinted with the permission of Cambridge University Press.*

The next analysis examines speakers' eye movements. Figure 6 shows the percent of this initiation time window that speakers spent looking at the object in Region 1 (e.g., the 4-legged cat). The results show that high spans spent less of their available time fixating Region 1 than low spans did if there was a contrast object in Region 3. WM did not correlate with this measure during control trials. This implies that high spans were less likely to be looking at Region 1 while preparing their utterances in the presence of a contrast.

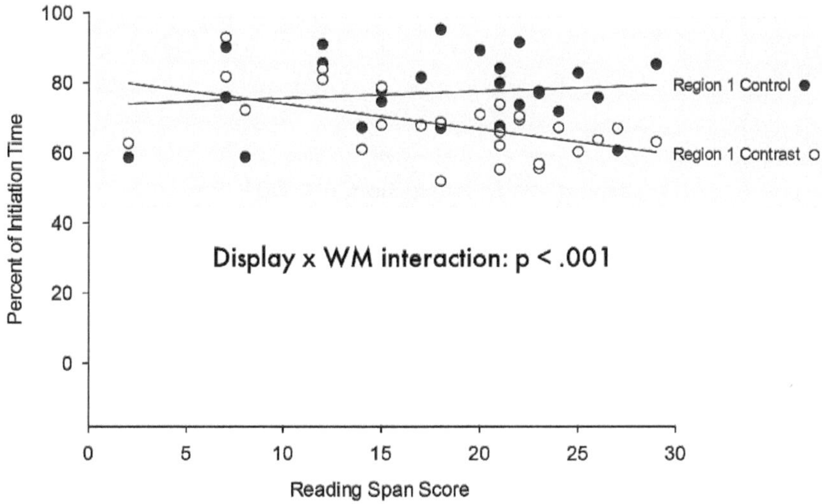

Figure 6. Gaze duration results from Swets et al. (2014). The period of interest in the displayed results is the time between the appearance of the stimulus, and the onset of speech. The visual region of interest is Region 1, the object occupying the first (left-most) position in the display. Reprinted from Language and Cognition, *6, Issue 01, 2014, 12–44, Swets, B., Jacovina, M.E., and Gerrig, R.J. Individual differences in the scope of speech planning: Evidence from eye movements. Copyright © 2014 UK Cognitive Linguistics Association. Reprinted with the permission of Cambridge University Press.*

The next analysis reveals that high span speakers spent this extra time that they otherwise would use to look at Region 1 by looking at the object in Region 3. Figure 7 shows the percent of initiation time that speakers spent looking at the object in Region 3 (e.g., the 3-legged cat in the contrast condition vs. the wheel in the control condition). In fact, during contrast trials, high span individuals were more likely to fixate the contrast objects in Region 3 prior to articulation than low spans. Neither high spans nor low spans tended to look at this region if there was no contrast to encode for description.

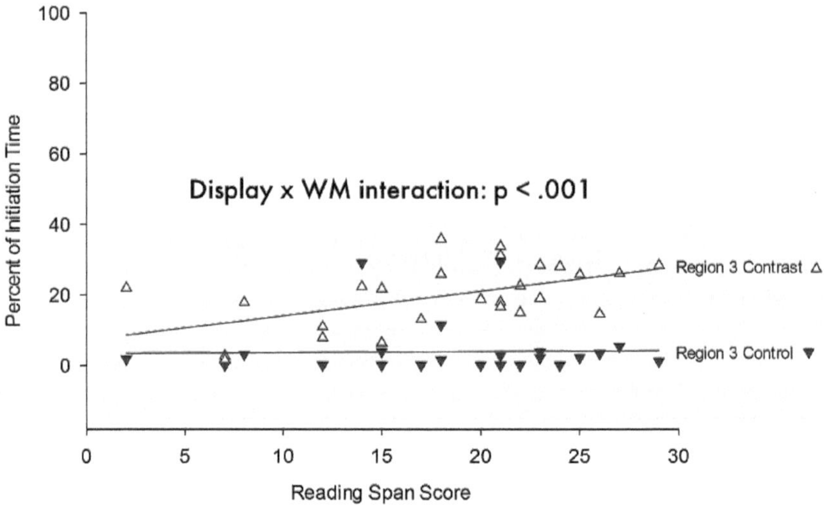

Figure 7. Gaze duration results from Swets et al. (2014). The period of interest in the displayed results is the time between the appearance of the stimulus, and the onset of speech. The visual region of interest is Region 3, the object occupying the third (right-most) position in the display. Reprinted from Language and Cognition, *6, Issue 01, 2014, 12–44, Swets, B., Jacovina, M.E., and Gerrig, R.J. Individual differences in the scope of speech planning: Evidence from eye movements. Copyright © 2014 UK Cognitive Linguistics Association. Reprinted with the permission of Cambridge University Press.*

The next analysis will reveal whether the speakers with more working memory who were more likely to literally look ahead to Region 3 of the display also produced correspondingly detailed descriptions. Speakers in general created longer descriptions in the presence of a contrast, and those with more working memory were much more likely to do so by modifying the first noun. Figure 8 presents the likelihood that a speaker modified N1, either by calling the first cat a *four-legged cat*, *cat with four legs*, *whole cat*, or *cat with two legs*. The figure shows that working memory predicted this likelihood such that high spans were more likely to modify N1 than low-spans if there was a contrast. There was no such correlation during control trials. This result suggests that high spans are not only more apt to gather information about the contrast prior to speech, but also to encode that contrast very early on into their utterance plans.

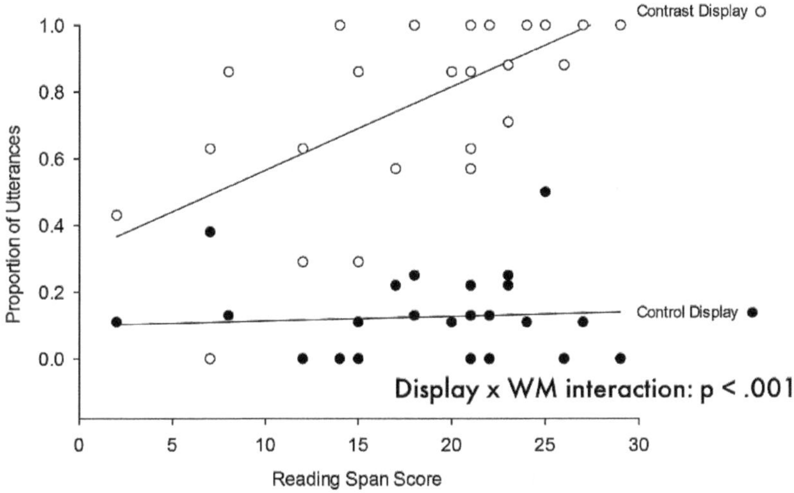

Figure 8. Utterance type results from Swets et al. (2014) showing the likelihood that participants modified the first noun phrase in a target utterance (N1). Reprinted from Language and Cognition, *6, Issue 01, 2014, 12–44, Swets, B., Jacovina, M.E., and Gerrig, R.J. Individual differences in the scope of speech planning: Evidence from eye movements. Copyright © 2014 UK Cognitive Linguistics Association. Reprinted with the permission of Cambridge University Press.*

To summarize, we found that reading span predicted speakers' scope of planning. One of the first signs of these individual differences appeared during the initiation time window. Although high spans took the same amount of time to begin speaking as low spans when a contrast was present, they used that time much differently than low spans. Specifically, high spans spent more time gathering information about the similar objects to be distinguished for the matcher. This additional inspection of the more advance regions in the display allowed them to integrate information about the contrast earlier than low spans: They not only gave longer descriptions of N1, but also showed a greater likelihood of modifying N1 to verbalize the contrast with N3.

In support of the general hypothesis, we found that higher working memory capacity is associated with a larger scope of speech planning. It appears that speakers with high verbal working memory capacity are able to not only gather more information about a message before speaking, but also integrate that message early on in utterance plans. On the other hand, low spans are not as productive in using the time available to gather advance planning information. These results thus favor the efficient capacity view of working memory's role in planning over the simple capacity view. Working memory capacity seems to allow speakers to be more efficient, or productive, in the extent to which they plan utterances in advance. High spans can not only plan more utterance information than low spans at a given time, but as Figure 5 shows, they created these larger plans without taking any additional time to do so. An account holding that WM simply stores larger utterance plans cannot account for such an effect. These effects are consistent with those observed by Petrone et al. (2011), who had found similar effects of efficient advance planning among individuals with high WM. Together these results argue for the role of working memory in sentence planning as a provider of efficient information integration in addition to a storage space.

Now that I have illustrated how the individual differences approach can help us to better understand the role of working memory in sentence production, I will now return to the domain of sentence comprehension. In doing so, I hope to illustrate that working memory's role in both is quite similar.

5. The role of working memory in language processing

At the outset of the paper, I presented results from a study (Swets et al., 2007) showing that individual differences in working memory accounted for more systematic variance in relative clause attachment preferences than cross-linguistic differences, which suggests that working memory is involved in sentence comprehension in substantive ways. Then, I demonstrated that working memory plays a role in sentence planning, and illustrated that the nature of this role entails more than simple storage. The aim of this section is to show that individual differences approaches applied to multiple sentence processing domains can help explain the role that working memory might play more generally.

Recall from the earlier study (Swets et al., 2007) that individual differences in working memory capacity predicted ambiguous relative clause (RC) attachment such that high-spans attached low and low-spans attached high. We had not predicted this result. Our predictions had been based on assumptions regarding the mechanistic role that WM must play in parsing. Specifically, we had supposed that WM plays the role of simple storage: The more WM one had, the more likely they were to keep NP1 available in storage long enough to be associated with the RC. Given that we found the opposite result, we were forced to examine an alternative view of the role that working memory plays in this process.

To explain this finding, we proposed that WM plays a role that exceeds that of simple storage. Rather, it is a processing resource that allows comprehenders to chunk a certain amount of information together while reading. If high-span readers can "chunk" more information together while reading, they can regard the entire subject of the sentence as one "processing unit". On the other hand, low-spans may have to break up the subject because of its length. A likely boundary for such a break point is just before the relative clause, which would separate the complex noun phrase ("The sister of the actress") from the relative clause ("who shot herself on the balcony"). By placing such a mental boundary at that point, the NP1 could become a more appealing attachment site than NP2. Hence, our hypothesis was that chunking strategies underlie the individual differences observed in Study 1: Perhaps the reason low-span readers attach to NP1 is that they create smaller "processing chunks" as they read silently, leading to NP1 being the more viable attachment site. If this is true, then forcing all readers (including high spans) to use the same chunking strategies during reading should reduce the attachment preference differences between high- and low-spans.

Study 2 tested this hypothesis by forcing participants to parcel the complex NP and the RC into two pieces with an intervening break. Specifically, we presented each of the Study 1 sentences in 2-second chunks: first, the complex NP ("The maid of the princess"), followed by the relative clause with a modifying prepositional phrase PP ("who shot herself on the balcony"), then the matrix verb phrase VP ("was under investigation"). According to this hypothesis, if WM underlies the size of the processing chunks people use to parse syntax, then forcing a break between N2 and the RC should reduce or eliminate the relationship between WM and attachment preference by making everyone behave like low spans, whereby they attach high, to NP1.

Figure 9 reveals that chunking the text had precisely this effect on relative clause attachment preferences. The left side of the graph shows the Study 1 effects that were summarized earlier, and illustrates that participants with lower working memory were more likely to attach "high", to NP1. The right side shows the results of Study 2, in which the text of the relative clause sentences was artificially chunked. The figure reveals that the relationship between Reading Span and RC attachment preferences was greatly reduced in English and apparently eliminated in Dutch. Also noteworthy is that percent NP1 attachments increased for all groups in both languages, and that English attachment and Dutch attachment both revealed an overall NP1 preference.

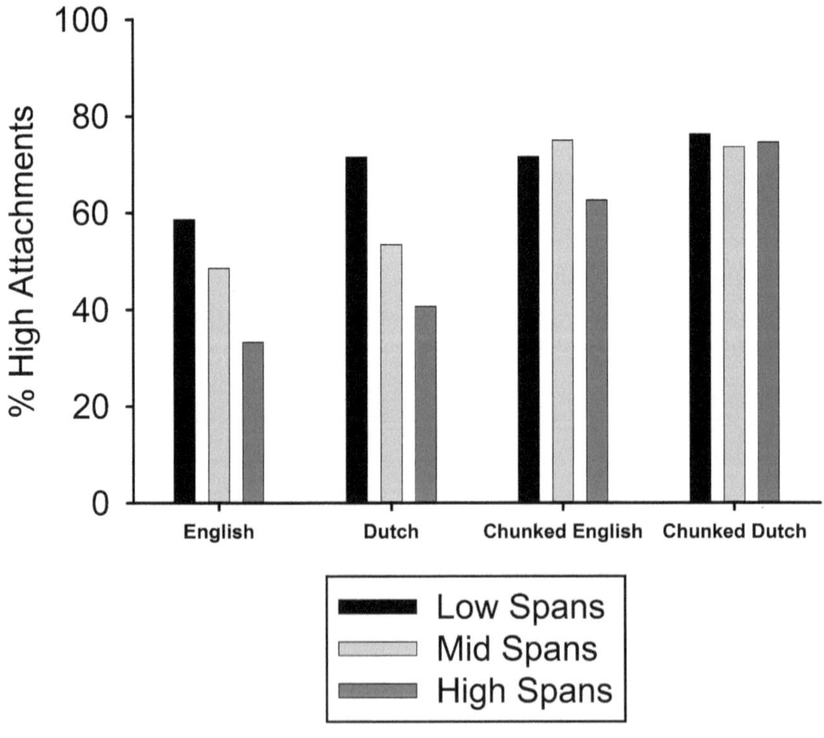

Figure 9. Overall results from Swets et al. (2007): Attachment preferences as a function of language (English vs. Dutch), presentation style (whole vs. chunked), and working memory category (low, mid and high span).

To sum up, in Study 1, the direction of the relationship between WM and attachment preference was the same in both English and Dutch: Individuals low in WM attached high, and individuals high in WM attached low. In Study 2, chunking the text reduced the relationship between WM and attachment preference significantly because it effectively forced everyone to adopt the chunking strategies used by low spans. From this set of results, we can draw two conclusions. First, the final products of parsing are bounded by the limits of working memory capacity. But more germane to the present argument, we can also conclude that the mechanistic role that working memory plays in this parsing process is not simply to store potential attachment sites. Rather, working memory is functioning at a high level during which initial packages of information are assembled together. The more working memory one has, the more likely one is to assemble large packages of information to be analyzed for later parsing decisions.

By examining this study in concert with the study on individual differences in planning scope, we can also draw a conclusion about the role of working memory in language processing more generally: Working memory predicts informational chunking not just in the manner in which we plan our sentences during language production (Swets et al., 2014), but also in aspects of language comprehension. Moreover, its role cannot be reduced to that of simple storage.

Working memory's role was similar in both domains: In the sentence planning study (Swets et al., 2014), participants were able to create larger utterance plans with additional WM; and in the attachment preferences study, participants were able to package more linguistic material together for parsing/analysis with additional WM. Taken together, this suggests that in perhaps all domains of sentence processing, working memory plays the role of packaging information together for both purposes of storage and active, efficient integration with other information. Furthermore, any differences that can occur in working memory capacity be they individual differences or experimentally manipulated reductions in capacity, by influencing the size of these packages and the efficiency with which they are assembled, can influence the basic mechanisms of sentence processing. I am currently collecting data from other domains to determine whether this principle of working memory as information packager applies even more generally. One such area of investigation is in the use of lexical and event information to predict upcoming elements in a sentence (Altmann and Kamide, 1999), which could show that the scope of prediction, like the scope of planning or the size of informational packages assembled during parsing, varies along with working memory.

6. Future directions

So far, I have summarized some previous research demonstrating that individual differences research can help explain variance in the scope of planning in language production. Although this early research is promising, there is still more work that can be done in this research vein to help understand the cognitive architecture associated with planning in language production. Such future directions of this research ought to include additional individual differences measures, examine more levels of planning, and compare individual differences to cross-linguistic differences in planning scope.

More individual differences measures. To this point, the only individual differences measures that have been shown to correlate with the scope of advance speech planning are working memory (Petrone et al., 2011; Swets et al., 2014) and preparation time (Gillespie and Pearlmutter, 2011; Lange and Laganaro, 2014; Wagner et al., 2010). But WM is known to correlate with other aspects of cognition such as attentional control (Kane et al., 2004) and processing speed (Salthouse, 1994). To complicate things even further, there are multiple aspects of working memory, including possible systems that serve strictly verbal WM, strictly spatial WM, and a more general WM that underlies all cognitive processing (Swets et al., 2007). The effects of WM on planning processes may represent modest advances, but they also do not account for enough variance to consider the case closed. In other words, there are more sources of individual differences to pursue, including for example age, attentional control, and personality factors.

Two sources of potential systematic variance that seem especially ripe for further investigation are processing speed and speech rate. It is possible that individual differences in processing speed might account for the observed link between working memory and advance sentence planning. Prior research (summarized in Salthouse, 1994) has documented that age-related declines in working memory can be largely attributed to declines in processing speed, which points toward a possible processing speed explanation of our WM findings. Furthermore, it is also possible that processing speed, the rate at which speakers articulate linguistic material, and planning might be interrelated in ways that to this point have not been explored in the literature. Hence, it will be important in future research to include measures of processing speed separable from working memory capacity to sort out how these various facets of cognitive performance help facilitate the speed, fluency, and scope of planning in language production.

Of course, the search for other individual differences factors that might be associated with the planning scope of language production need not be limited to just the above factors. For example, in other domains of language production such as syntactic priming, measures of individual differences including Big Five personality factors such as extraversion and conscientiousness (Gill et al., 2004), age (Kidd, 2012), and perspective-taking and autism (Horton, 2014) have been found to correlate with the extent to which a speaker will re-use a syntactic structure just uttered by an interlocutor. Perhaps such factors could explain meaningful variance in planning scope as well. For example, Horton (2014) had found that individuals higher in perspective-taking ability were more likely to align with their partner on the type of syntactic structure they produced in a picture description task. In other words, individuals who had been measured to be highly sensitive to their partner's perspectives produced language that also seemed sensitive to the perspectives of their partners. Given that an essential element of proper advance planning in the picture description task described in Swets et al. (2014) is understanding the communicative needs of the addressee, it is possible that individuals higher in perspective-taking would also be more likely to plan more of their sentence in advance of articulation.

One other individual differences measure that deserves some mention in this context is something known as the BLIRT measure (Brief Loquaciousness and Interpersonal Responsiveness Test, Swann and Rentfrow, 2001). "Blirtatiousness", defined by Swann and Rentfrow (2001) as "how quickly, frequently, and effusively people respond to their partners" could be used as a measure to variance related to personality, perspective-taking, and temporal factors related to the planning of language. Those high in "blirtatiousness" talk quickly and often, and those low in the factor are more measured in their speech output. Use of this scale in future research might capture a great deal of variance in planning scope among speakers, but also risks being too general a measure to explain specific mechanisms.

Levels of planning. An additional limitation of the research so far is that measures of WM have only been shown to correlate with two aspects of speech planning: prosodic plans (Petrone et al., 2011) and the interface of message level planning and utterance planning (Swets et al., 2014). But limitations of this previous research invite more work to be done. For example, one limitation of the research on individual differences in prosodic planning (Fuchs et al., 2013; Petrone et al., 2011) is that it is based on read speech. However, in line with Swets et al. (2014), it would be preferable to examine speech planning in more interactive situations, when the linguistic content is generated primarily by the speaker.

But the primary limitation of this previous research is that it has considered only a small sample of the range of representations that are planned during language production. As mentioned above, speech must be planned at several levels of representation (Levelt, 1989). With only message level and prosodic level representations considered so far in the study of individual differences in planning scope (Petrone et al., 2011, Swets et al., 2014), other levels to be considered include phonological encoding, grammatical encoding, and lemma selection. Hence, future projects ought to examine the planning scope of language production at multiple levels of representation as a function of multiple measures of individual differences.

A cross-linguistic approach. Similar to the research line taken in Swets et al. (2007), one intriguing direction of individual differences research in the domain of language production might be to compare effects of individual differences on speech planning across multiple languages. Although some previous research (Janssen, et al., 2008; Brown-Schmidt and Konopka, 2008, Christianson and Ferreira, 2005; Schriefers and Teruel, 1999b; see Jaeger and Norcliffe, 2009, for a review) has investigated differences in incremental planning between languages, and some more recent research has focused on individual differences, no previous research on the flexibility of planning scope has ever simultaneously compared planning scope differences among various languages to planning scope differences among individuals of the same language community. It would be interesting to examine how the cross-linguistic differences in grammatical encoding scope found by Schriefers and Teruel (1999b), for example, would compare in effect size to individual differences in the scope that is predicted by WM. Cataloging the factors that influence planning scope within and across languages can help researchers map the contours of the language production system that all languages share.

Notes of caution. I hope that this chapter conveys the enthusiasm I have for individual differences approaches in studies of speech production and perception. But before I conclude, it seems appropriate to give some words of caution regarding their utility. First, let me repeat the most important warning: Because individual differences research is inherently correlational, one cannot claim a causal link between an individual differences measure and performance on some linguistic task. Fortunately, there are ways to address this limitation of the approach. First, although one cannot assume a causal link when finding a significant correlational relationship, correlation also does not *rule out* the existence of such a link. So perhaps it is sometimes better to regard a significant correlation between, say, working memory and planning scope, not as evidence that increased capacity causes increased planning scope, but as initial evidence for such a link which must be confirmed by subsequent research. To find support for such a causal link, one could conduct experiments by manipulating a task circumstance that is thought to use resources that are associated with the individual differences measure that has already been shown to correlate with a linguistic measure of interest. For example, if we are considering working memory, manipulation of task load as in Wagner et al. (2010) can simulate the high- or low-span functionality that is naturally expressed in individual differences. A more exploratory, but intriguing technique to help confirm causal explanations of individual differences findings is tDCS (trans-cranial direct current stimulation). With tDCS, a researcher places anode and cathode electrodes at different sites on the scalp and delivers a low-voltage current through the head. By placing the anode electrode over a region of interest, one can "stimulate" that region. One way this technique has been used in language studies has been to show that placing an anodal electrode over the prefrontal cortex (PFC), associated with executive function, participants planned sentences that were less error-prone than participants who received a control (sham) stimulation condition (Nozari, et al., 2014). Grammatical fluency in conversational speech has also improved by placing the anode over Broca's area (Marangolo et al., 2013). One appealing aspect of tDCS over task manipulation in general is that one can attempt to isolate *and improve* the cognitive process (WM, executive function, grammar) that is thought to explain the phenomenon under investigation rather than develop some task that merely disrupts task performance.

Beyond the interpretational limitations of correlational research, studies of individual differences tend to require larger numbers of participants than traditional experimental designs. The number required follows from the kinds of questions one wants to address. For example, in our study on individual differences in relative clause attachment (Swets et al., 2007), we were testing hypotheses for which we needed to tease apart different sources of working memory variance to see how they each predicted relative clause attachment preferences. To do this, we required two types of statistical analyses that are quite greedy regarding numbers of participants: factor analysis and structural equation modeling. Our large samples ranging from 96 to 150 participants per study proved to be manageable in a study of sentence comprehension with easily codable binary responses, but for studies of language production with their large demands on time for transcription and coding, such numbers of participants can prove to be intractable. Our dual-solution to this problem in the working memory/planning scope research (Swets et al., 2014) was to perform simpler statistical analyses using mixed models in R, and to "rig" the working memory sample to ensure greater variability. Although we initially collected working memory data from nearly 100 participants, we invited only 26 participants back to participate as speakers in the picture description task. The 26 who returned showed a wide range of working memory scores because we invited back everyone who scored in the more extreme ends of the distribution, and sent fewer invitations to participants in the center. By maximizing the variance in working memory, we were able to find a significant relationship between planning scope and working memory among this small number of participants.

This final warning is one that would seem obvious, but still warrants a mention. Of course every researcher understands that it is inadvisable to conduct unmotivated studies. However, this advice is far easier to heed when the studies one conducts involves experimentally manipulated variables that one must counterbalance and fuss over before any data are collected. The temptation when dealing with individual differences measures is to include them in any study one performs, without necessarily justifying their inclusion. Although individual differences measures have great potential to bring language production researchers an increased understanding of variables related to speech planning, we must be judicious in using them, and provide sound theoretical motivation for every measure we take.

References

Altmann, G. T. M. and Kamide, Y. (1999). Incremental interpretation at verbs: Restricting the domain of subsequent reference. *Cognition*, *73*, 247–264.

Ashcraft, M. J. (1992). Cognitive arithmetic: A review of data and theory. *Cognition*, *44*, 75–106.

Bock, J. K., and Levelt, W. J. M. (1994). Language production: Grammatical encoding. In M. A. Gernsbacher (Ed.), *Handbook of psycholinguistics* (pp. 945–984). San Diego: Academic Press.

Brown-Schmidt, S. and Konopka, A. E. (2008). Little houses and casas pequeñas: message formulation and syntactic form in unscripted speech with speakers of English and Spanish. *Cognition*, *109*, 274–280.

Brown-Schmidt, S., and Tanenhaus, M. (2006). Watching the eyes when talking about size: An investigation of message formulation and utterance planning. *Journal of Memory and Language*, *54*, 592–609.

Brysbaert, M., and Mitchell, D. C. (1996). Modifier attachment in sentences parsing: Evidence from Dutch. *Quarterly Journal of Experimental Psychology, 49,* 664–695.

Christianson, K. and Ferreira, F. (2005). Planning in sentence production: Evidence from a free word-order language (Odawa). *Cognition*, *98*, 105–135.

Costa, A., and Caramazza, A. (2002). The production of noun phrases in English and Spanish: Implications for the scope of phonological encoding in speech production. *Journal of Memory and Language*, *46*, 178–198.

Cuetos, F., and Mitchell, D. C. (1988). Cross-linguistic differences in parsing: restrictions on the use of the Late Closure strategy in Spanish. *Cognition, 30,* 73–105.

Daneman, M., and Carpenter, P. A. (1980). Individual differences in working memory and reading. *Journal of Verbal Learning and Verbal Behavior*, *19*, 450–466.

Damian, M. F., and Dumay, N. (2007). Time pressure and phonological advance planning in spoken production. *Journal of Memory and Language*, *57*, 195–209.

Ferreira, F., and Swets, B. (2002). How incremental is language production? Evidence from the production of utterances requiring the computation of arithmetic sums. *Journal of Memory & Language*, *46*, 57–84.

Ferreira, F., and Swets, B. (2005). The production and comprehension of resumptive pronouns in relative clause 'island' contexts. In: Cutler, A. (Ed.), *Twenty-first Century Psycholinguistics: Four Cornerstones* (pp. 263–278). New Jersey: Lawrence Erlbaum Associates Publishers.

Ford, M., and Holmes, V. M. (1978). Planning units and syntax in sentence production. *Cognition*, 6, 35–53.

Frazier, L. (1979). On Comprehending Sentences: Syntactic Parsing Strategies. Doctoral dissertation, University of Connecticut.

Frazier, L. (1987). Theories of sentence processing. In L. J. Garfield (Ed.), *Modularity in knowledge representation and natural-language understanding*. Cambridge, MA: MIT Press.

Fuchs, S., Petrone, C., Krivokapic, J., and Hoole, P. (2013). Acoustic and respiratory evidence for utterance planning in German. *Journal of Phonetics*, *41*, 29–47.

Garrod, S., and Pickering, M. J. (2007). Automaticity of language production in monologue and dialogue. In A. S. Meyer, L. R. Wheeldon, A. Krott, (Eds.), *Automaticity and control in language processing* (pp. 1–20). New York, NY US: Psychology Press.

Gill, A. J., Harrison, A. J., and Oberlander, J. (2004). Interpersonality: Individual differences and interpersonal priming. *Proceedings of the 26th Annual Meeting of the Cognitive Science Society* (pp. 464–469). Mahwah, NJ: Erlbaum.

Gillespie, M. and Pearlmutter, N.J. (2011). Effects of semantic integration and advance planning on grammatical encoding in sentence production. In L. Carlson, C. Holscher, T. Shipley (Eds.), *Proceedings of the 33rd Annual Conference of the Cognitive Science Society* (pp. 1625–1630). Austin, TX: Cognitive Science Society.

Griffin, Z. M. (2001). Gaze durations during speech reflect word selection and phonological encoding. *Cognition*, *82*, B1–B14.

Hartsuiker, R. J., and Barkhuysen, P. N. (2006). Language production and working memory: The case of subject-verb agreement. *Language and Cognitive Processes, 21*, 181–204.

Horton, W. S. (2014). Individual differences in perspective taking and field-independence mediate structural persistence in dialog. *Acta Psychologica, 150*, 41–48.

Horton, W. S., and Spieler, D. H. (2007). Age-related differences in communication and audience design. *Psychology and Aging*, 22, 281–290.

Jaeger, T. F. and Norcliffe, E. (2009). The cross-linguistic study of sentence production. *Language and Linguistics Compass*, *3*, 866–887.

Janssen, N., Alario, F.-X, and Caramazza, A. (2008). A word-order constraint on phonological activation. *Psychological Science*, *19*, 216–220.

Kane, M. J., Hambrick, D. Z., Tuholski, S. W., Wilhelm, O., Payne, T. W., and Engle, R. W. (2004). The generality of working memory capacity: A latent-variable approach to verbal and visuospatial memory span and reasoning. *Journal of Experimental Psychology: General*, *133*, 189–217.

Kellogg, R. T., Oliver, T., and Piolat, A. (2007). Verbal, visual, and spatial working memory in written language production. *Acta Psychologica*, *124*, 382–397.

Kemper, S., Herman, R. E., and Lian, C. H. T. (2003). The costs of doing two things at once for young and older adults: Talking while walking, finger tapping, and ignoring speech or noise. *Psychology and Aging*, *18*, 181–192.

Kemper, S., and Sumner, A. (2001). The structure of verbal abilities in young and older adults. *Psychology and Aging*, *16*, 312–322.

Kidd, E. (2012). Individual differences in syntactic priming in language acquisition. *Applied Psycholinguistics*, *33*, 393–418.

Konopka, A. E., and Meyer, A. S. (2010). Looking ahead: Variability in planning scope for complex noun phrases – evidence from eye-tracking [Abstract]. *In Proceedings of the 16th Annual Conference on Architectures and Mechanisms for Language Processing [AMLaP 2010]* (pp. 177). York: University of York.

Korvorst, M., Roelofs, A., and Levelt, W. J. M. (2006). Incrementality in naming and reading complex numerals: Evidence from eyetracking. *The Quarterly Journal of Experimental Psychology*, *59*, 296–311.

Lange, M.V., and Laganaro, M. (2014). Inter-subject variability modulates phonological advance planning in the production of adjective-noun phrases. *Frontiers in Psychology*, *5*, 43.

Levelt, W. J. M. (1989). *Speaking: From intention to articulation*. Cambridge, MA: MIT Press.

Marangolo, P., Fiori, V., Calpagnano, M. A., Campana, S., Razzano, C., Caltagirone, C., and Marini, A. (2013). tDCS Over the left inferior frontal cortex improves speech production in aphasia. *Frontiers in Human Neuroscience*, *7*, 539.

Nozari, N., Arnold, J. E., and Thompson-Schill., S. L. (2014). The effects of anodal stimulation of the left prefrontal cortex on sentence production. *Brain Stimulation*, *7(6)*, 784–792.

Petrone, C., Fuchs, S., and Krivokapić, J. (2011). Consequences of working memory differences and phrasal length on pause duration and fundamental frequency. Paper presented at the Proceedings of the 9th International Seminar on Speech Production (ISSP), 393–400. Montréal, Canada.

Salthouse, T. A. (1994). The aging of working memory. *Neuropsychology*, *8*, 535–543.

Schriefers, H., and Teruel, E. (1999a). Phonological facilitation in the production of two-word utterances. *European Journal of Cognitive Psychology*, *11*, 17–50.

Schriefers, H., and Teruel, E. (1999b). The production of noun phrases: a cross-linguistic comparison of French and German. In *Proceedings of the Twenty-First Annual Conference of the Cognitive Science Society*, (Mahwah, NJ: Lawrence Erlbaum), 637–642.

Schriefers, H., Teruel, E. and Meinshausen, R. M. (1998). Producing simple sentences: results from picture-word interference experiments. *Journal of Memory & Language*, *39*, 609–632.

Slevc, L. (2011). Saying what's on your mind: Working memory effects on sentence production. *Journal of Experimental Psychology: Learning, Memory, and Cognition*, *37*, 1503–1514.

Smith, M., and Wheeldon, L. (1999). High level processing scope in spoken sentence production. *Cognition*, *73*, 205–246.

Smith, M., and Wheeldon, L. (2001). Syntactic priming in spoken sentence production—an online study. *Cognition*, *78*, 123–164.

Swann, W. B. Jr. and Rentfrow, P. J. (2001). Blirtatiousness: Cognitive, behavioral, and physiological consequences of rapid responding. *Journal of Personality and Social Psychology*. 6, 1160–1175.

Swets, B., Desmet, T., Hambrick, D. Z., and Ferreira, F. (2007). The role of working memory in syntactic ambiguity resolution: A psychometric approach. *Journal of Experimental Psychology: General*, *136*, 64–81.

Swets, B., Jacovina, M. E., and Gerrig, R. J. (2013). Effects of conversational pressures on speech planning. *Discourse Processes*, *50*, 23–51.

Swets, B., Jacovina, M. E., and Gerrig, R. J. (2014). Individual differences in the scope of speech planning: Evidence from eye movements. *Language and Cognition*, 6, 12–44.

Wagner, V., Jescheniak, J. D., and Schriefers, H. (2010). On the flexibility of grammatical advance planning during sentence production: Effects of cognitive load on multiple lexical access. *Journal of Experimental Psychology: Learning, Memory, and Cognition*, *36*, 423–440.

Francesco Cangemi, Martina Krüger and Martine Grice
Universität zu Köln, IfL Phonetik

Listener-Specific Perception of Speaker-Specific Productions in Intonation

We see earth by earth, water by water
Bright aether by aether, and obliterating fire by fire
Love by love, and strife by baneful strife
Empedocles, Fragment 109

Abstract: In this contribution we explore the hypothesis of an interaction between speaker- and listener-specific strategies in the encoding and decoding of intonational contrasts. Intonational categories, such as the pitch accents used in the signalling of focus types, can be cued by different phonetic exponents, such as peak alignment or duration of the target words. Through a production task we document speaker-specific strategies: Individual speakers might use more or fewer cues than others (robustness) when encoding intonational contrasts, and each cue can be used to encode one or more contrasts (partitioning). We show in a subsequent perception task that listeners are sensitive to speaker-specific strategies, since correct identification scores for productions of individual speakers mirror the robustness and partitioning of speakers' productions. Moreover, listeners vary as to how reliably they decode intonational contrasts across speakers. However, in line with the hypothesis of an interaction between speaker- and listener-specific behaviours, some listeners are more reliable at decoding contrasts as encoded by some particular speakers, which in turn are decoded less reliably by other listeners. These findings suggest that phonetic cues to intonational contrasts should not be understood as singly necessary and jointly sufficient features for category membership, but rather as dimensions along which phonological categories cluster, in an individual-specific network of phonological knowledge.

1. Introduction

1.1. Background

It is not an overstatement to say that in recent years phonetic research has stepped away from the brutal averaging of data points collected across subjects, be it in perception or production, although this has only just begun to play a role in research into intonation. We begin by taking the example of the voicing contrast in plosives, particularly in syllable initial position. In their seminal paper, Lisker and Abramson (1964) collected acoustic data from 11 languages, represented by only 17 speakers altogether. Despite the fact that their four speakers of American English showed massive differences in their individual behaviour (Lisker and Abramson, 1964: 538), voicing in Dutch, Tamil, Cantonese, Eastern Armenian, Korean, Hindi and Marathi was investigated using data from a single speaker for each language. Later on, in his groundbreaking study on the phonetic exponents of voicing in word-internal stops, Lisker (1986) reviewed 16 acoustic cues to voicing, again under the tacit assumption that the weighting of such cues would not be affected by listener-specific patterns. Individual specificity in production and perception was thus out of the picture in these two studies, which focussed on cross-language comparisons and on the relationships between articulation and perception, respectively.

In recent years, however, speaker- and listener-specific behaviour has gained a central role in the study of how phonetic substance maps onto phonological contrasts. This evolution might have stemmed from the ability of linguists to integrate insights and practices from neighbouring disciplines, both at the theoretical level (as in the case of a renewed understanding of category structure, e.g. Lakoff, 1987) and at the methodological level (as with mixed-effects modelling, notably through the targeted exploration of random coefficients, e.g. Baayen, 2008). As a consequence, recent studies on voicing contrasts in stops have devoted a great deal of attention to speaker- and listener-specific behaviour – not only as important factors in the data analysis, but also as dimensions shaping the actual research questions. Allen et al. (2003), for example, document systematic variation of voice onset time patterns in stop contrasts across speakers, and link this finding to speaker recognition mechanisms. Individual differences are found in the weighting of the cues associated with stop contrasts in production (e.g. Schultz et al., 2012, for voicing in English) and perception (e.g. Idemaru et al., 2012, for stop length in Japanese). Research on individual behaviour has also been conducted in the effort to provide evidence in favour of theories suggesting a link between production and perception. Recent studies in this vein include Perkell et al. (2004a, b), which tested the hypothesis that the more precisely a subject discriminates a contrast as a listener, the more accurately that subject will produce such contrast as a speaker, both in terms of articulation patterns and acoustic output. Findings from Bradlow et al. (1996), Newman et al. (2001), Hazan and Baker (2011) and Hazan et al. (2013) are also compatible with the assumption of more accurate production resulting in greater intelligibility. Speaker and listener-specific behaviours are thus well attested for segmental contrasts, which have been studied extensively in the past fifty years.

The situation for intonation (and prosody in general) is radically different. Despite the fact that there is an abundance of studies reporting on language-specific marking of focus types using accent types, deaccentuation or dephrasing (e.g. Jun, 2014), only few studies focussed on individual-specific differences, notably in production. An unpublished study by Andreeva and Barry (2007) on phrasal prominence suggests that its realization differs not only across the investigated languages (Bulgarian and Russian), but also between the speakers of each of the two languages. Niebuhr et al. (2011) show that an intonational contrast in Standard Northern German is cued by one group of speakers through differences in peak alignment, and by another group through differences in peak shape. This paper did not investigate the consequences of these different production strategies for perception. In fact, to our knowledge, no studies have targeted listener-specific strategies in the decoding of intonational contrasts – let alone the interaction between specificity in production and perception.

1.2. Rationale

In this contribution, we explore the *interaction* between speaker- and listener-specific behaviour in the encoding and decoding of prosodic categories. Note that this is different from exploring the *link* between speaker- and listener-specific behaviour, as in the studies by Perkell et al. (2004a,b) cited above, in which subjects participated in both a production and a perception task. Their results showed that some individuals produce contrasts more accurately than others, that some individuals discriminate contrasts more precisely than others, and that accurate speakers are also precise listeners (see Figure 1).

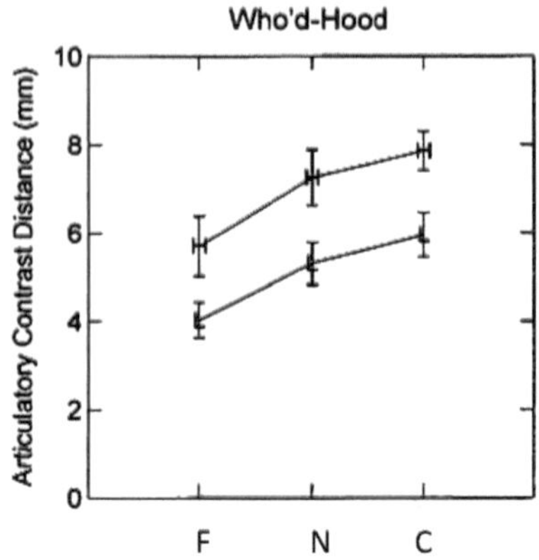

Figure 1: Articulatory contrast distance for tongue body position (y-axis; error bars are one standard error about the mean) as a function of three speaking conditions (x-axis; Fast, Normal, Clear) for the /ʊ, u/ contrast. Subjects are split into two groups based on their performance in a two-step discrimination task involving stimuli on the [ʊ, u] *continuum. Listeners are in the "high discrimination ability" (H) group if their responses are 100% correct, otherwise they are in the "low discrimination ability" (L) group. Reprinted from* Journal of Acoustical Society of America *116, 2338–2344. Perkell, J., Guenther, F., Lane, H., Matthies, M., Stockmann, E., Tiede, M., and Zandipour, M., The distinctness of speakers' productions of vowel contrasts is related to their discrimination of the contrasts. Reproduced with permission from AIP Publishing LLC. Copyright 2014.*

This methodology is particularly suited for documenting how good individuals are at producing and perceiving contrasts – that is, at profiling the "best speakers" and "best listeners", somehow assuming that there is a phonetic equivalent of the blood-type notions of "universal donors" and "universal receivers". Our aim is to show that not only some speakers might be generally more accurate and thus more intelligible than others, but also that some speakers might produce contrasts in a way that make them easily intelligible *to some particular listeners*, but not to others. This would document an interaction, rather than a link, between specificity in production and perception.

Figure 2 sums up the three potential scenarios. The behaviour of speakers and listeners could be independent, it could be linked or it could interact. In each panel, individuals are represented by nodes. Listeners are represented by empty circles identified by numbers and speakers are represented by filled circles identified by letters. The association lines between nodes represent how reliably a contrast produced by a given speaker is perceived by a given listener. Thick lines indicate that the intended categories produced by the speakers are frequently perceived correctly by the listener; thin lines indicate that this is rarely the case.

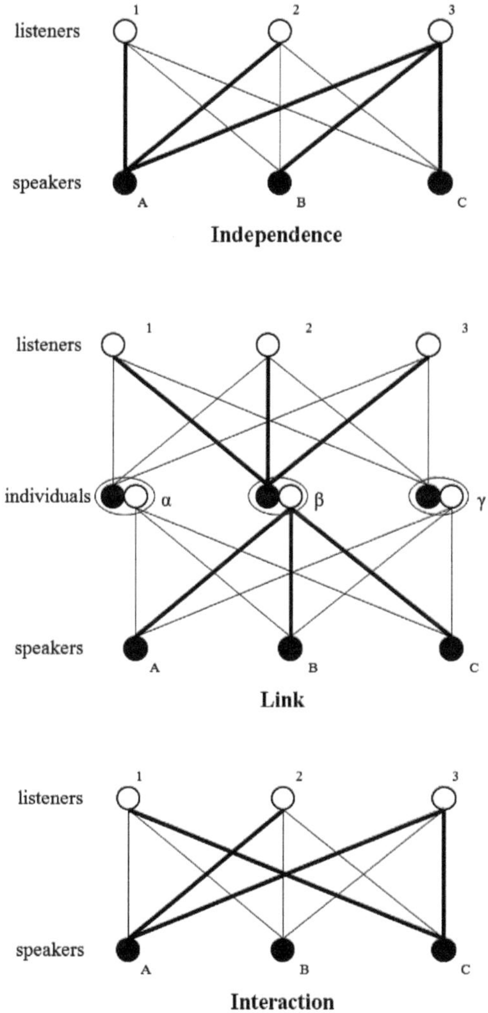

Figure 2: Independent, linked and interacting speaker- and listener-specific behaviour.

The top panel illustrates a situation in which some speakers (filled circles) are overall more intelligible and some listeners (empty circles) are overall more reliable, as indicated by the number of thick lines departing from each node. Crucially, the two phenomena are *independent*. Speaker A's productions are reliably perceived by all listeners, whereas listener 3 reliably perceives productions from all speakers. Using the blood types metaphor introduced above, speaker A would thus be an example of a "universal donor" and listener 3 would be a "universal recipient".

The mid panel depicts a *link* between production and perception within individuals. This is akin to the results reported by Perkell et al. (2004a,b), in which subjects participated in both a production and a perception task. The mid panel thus features three tiers, since subjects (in the central tier) serve as both listeners and speakers (hence the juxtaposed filled and empty circles identified by Greek letters), and thus both listeners (top tier) and speakers (bottom tier) are required. The illustration shows that some individuals (i.e. node β) are both accurate in their productions (thick lines connecting to listeners in the top tier) and reliable in their perceptual judgements (thick lines connecting to speakers in the bottom tier). Using the blood types metaphor introduced above, the individual β would thus be an example of an individual who is at the same time a "universal donor" and a "universal recipient".

The bottom panel illustrates the presence of an *interaction* between speaker- and listener-specific behaviour. Some speakers might still be overall more intelligible than others, and the same might apply for listeners. Crucially, however, there is no such thing as a "universally intelligible speaker" or a "universally proficient listener", as in the *independence* scenario, and thus (a fortiori) no individual who is both, as in the *link* scenario. Rather, a listener might perceive more reliably the contrasts produced by a given speaker, whose productions are in turn perceived less reliably by a different listener. Similarly, a given listener might be very reliable at decoding productions from a particular speaker, but perform very badly on productions from a different speaker. This is exemplified by speaker A being badly perceived by listener 1, who however is very reliable at decoding contrasts produced by speaker C. This is still compatible with some individual being overall better listeners (e.g. listener 3, with two thick lines departing from its node) or worse speakers (e.g. speaker B, with no thick lines departing from its node), as in the *independence* and *link* scenarios. However, in the *interaction* scenario these main effects can be modulated by specific interactions, and thus neither accuracy in production nor precision in perception need to be understood in absolute terms.

In the following sections, we explore the hypothesis of an interaction between speaker- and listener-specific strategies, using a dataset on the production and perception of focus in German collected for various purposes (Mücke and Grice, 2014, focussing on production; Grice et al., ms., focussing on perception). Before providing an analysis of the interaction between specific speaker and listener behaviours (3.3.), we thus summarise some of the relevant aspects of the two original studies.

2. Methods

2.1. Production task

Participants and recordings: Recordings were made of five speakers (three female) of Standard German from north of the Benrather isogloss, aged between 22 and 37 years. Articulatory movements were captured with a 2D Electromagnetic Articulograph (Carstens AG 100), with sensors on the upper and lower lips, recorded at 500 Hz, downsampled to 200 Hz and smoothed with a 40 Hz low-pass filter. Simultaneous acoustic recordings were made with a DAT-recorder (TASCAM DA-P1) using a condenser microphone (AKG C420 head set) and sampled at 44.1 kHz, 16 bit.

Materials: The materials contained target words /ˈbi:bɐ/, /ˈba:bɐ/, and /ˈbo:bɐ/ (fictitious names). These names were in the default position for the nuclear pitch accent (the last argument of the verb). Information structure was manipulated by means of question-answer pairs. Four different focus structures were elicited: the target word occurred either as part of the background or in broad, narrow or contrastive focus. An example of a set of question-answer pairs is given in Figure 3 for the target word <Bahber> /ˈba:bɐ/.

Questions:

1. Will Norbert Dr. Bahber treffen? *Does Norbert want to meet Dr. Bahber?*
2. Was gibt's Neues? *What's new?*
3. Wen will Melanie treffen? *Whom does Melanie want to meet?*
4. Will Melanie Dr. Werner treffen? *Does Melanie want to meet Dr. Werner?*

	Answers:	**test word in:**
	Melanie will Dr. Bahber treffen.	
1.	[———]focus	background
2.	[————————————]focus	broad focus
3.	[———]focus (Dr. Bahber)	narrow focus
4.	[———]focus (Dr. Bahber)	contrastive focus

(lit.: *Melanie wants Dr. Bahber to-meet*)

Figure 3: Speech material example, target word <Bahber> /ˈba:bɐ/. *Reprinted from* Journal of Phonetics *44, 47–61. Mücke, D. and Grice, M., The effect of focus marking on supra-laryngeal articulation – is it mediated by accentuation? Reproduced with permission from Academic Press.*

Subjects were presented with the contextualizing question both auditorily and visually. They then read aloud the answer in a contextually appropriate manner at a speaking rate which they considered to be normal. Question-answer pairs were randomized to avoid repetitions in sequences. In total, 560 tokens were recorded (4 target words x 4 focus structures x 7 repetitions x 5 speakers), although only 420 tokens were analysed (one target word having been discarded, owing to difficulties identifying lip aperture in the articulatory analysis).

Labels and measurements: Intonation was transcribed by two annotators using the acoustic waveform and F0 contours in PRAAT (Boersma and Weenink, 2010). In cases where transcribers differed (16%), a consensus transcription was reached. Accented target words were labelled using one of three different GToBI accent types (Grice et al., 2005): H+!H*, H* and L+H*, as presented schematically in Figure 4a-c. In all cases there was a low boundary tone sequence following, labelled as L-% (equivalent to L-L% in ToBI for English).

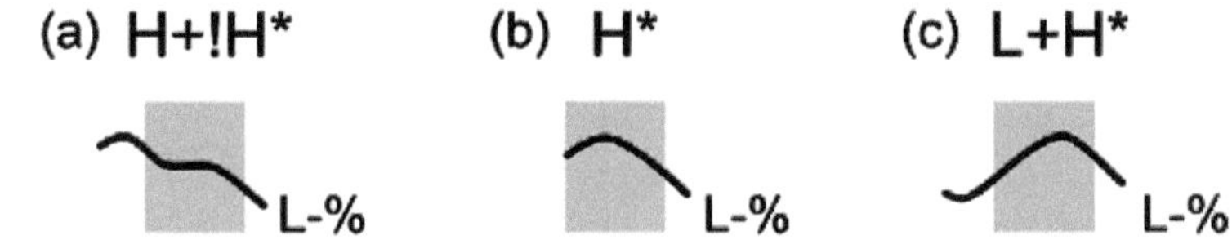

Figure 4: Schematic representation of the three different pitch accent types as presented in the GToBI online guidelines http://www.gtobi.uni-koeln.de.

Acoustic durations (target words and stressed syllables) were labelled by hand using the EMU speech database system (Cassidy and Harrington, 2001). For the kinematic recordings, the lip aperture index (LA, Byrd, 2000: 6) was calculated in terms of the Euclidean distance between the two sensors on the upper and lower lips capturing movements both in the horizontal and vertical dimensions. Kinematic labels were identified corresponding to the lip opening gesture in the stressed syllable, i. e. the movement from the maximum lip closure in the onset consonant to the maximum opening of the lips in the vowel, and the point of maximum velocity.

2.2. Perception task

Participants: Twenty native speakers of German (20 to 40 years of age, mean 25 years 9 months) with no knowledge of linguistics participated in the perception experiment. Participants had self-reported normal hearing.

Materials: Test sentences were taken from the production study described above. Thus, the carrier sentence "Melanie will Doktor _____ treffen." (Melanie wants to meet Doctor _____) contained one of the three fictitious target names: Bieber, Bahber and Bohber from one of the five speakers in one of the four focus conditions. Experimental materials contained the 420 tokens from the production study (3 target words x 5 speakers x 4 focus structures x 7 repetitions) plus 60 stimuli from a practice phase (4 focus structures x 3 target words x 5 speakers).

Every speaker was evaluated in a separate block, and within each speaker-block, every target name (Bieber, Bahber, Bohber) was evaluated separately. Target word blocks were randomized within the speaker blocks, and the speaker blocks were also randomized for each participant (controlled permutation). This allowed a controlled order of speaker blocks that was counterbalanced in order to avoid a possible influence on the judgments.

A practice phase of twelve stimuli (4 focus structures x 3 target words) preceded each speaker-block in order for the participants to familiarize themselves with the procedure and with possible speaker-specific strategies. For this practice phase, 12 prototypical stimuli were selected for each speaker, each target word and each focus structure. These items were those consistently assigned to the correct question/focus structure by six trained phoneticians in a pretest. In order to minimize learning effects, a given target word was only included in a single focus condition for each speaker. Practice phase stimuli were excluded from further analysis.

Procedure: The experiment was conducted with the PARADIGM software (Perception Research Systems 2007). Instructions were given in written form. The task was to match the test sentences heard to one of four questions (see Figure 3) presented on the screen. This was done by clicking on the question that subjects judged to be the most appropriate for a particular test sentence. There was no time limit for the choice.

In order to assure the comprehension of the task, participants were asked in a pretest to produce the target sentence (Melanie will Dr. Bahber treffen) as an answer to the four questions asked by the experimenter. None of the subjects reported difficulties in carrying out the task. Participants heard every test sentence once via headphones. The test sentences were preceded by a beep in order to assure full attention.

3. Results

3.1. Production

Table 1 shows a synopsis of the acoustic analysis, split by cues (rows) and speakers (columns); results refer to the three focus types (Broad, Narrow and Contrastive). Each cell shows how a given speaker uses a given cue in the encoding of the focus types; the tilde indicates absence of statistically significant differences between focus types[1]. Cells are displayed in different

1 Significance at $p = 0.05$ was assessed through ANOVAs run separately for each speaker and cue, and followed by post-hoc Tukey's HSD tests. For details on the

shades of grey according to the number of contrasts between focus types that a given cue allows for a given speaker. For example, peak height is significantly different for the three focus conditions in productions from speaker F3 (dark grey); for speaker M1, peak height is only significantly different in Broad focus cases, compared to both Narrow and Contrastive focus (light grey); for speaker M2, peak height does not vary across the three focus conditions (white). Speaker-specific differences are evident in both terms of *robustness*, involving the number of cues used in the encoding of the three categories, and in terms of *partitioning*, that is, whether a given cue is used to distinguish between two or more categories.

In Table 1, the number of white cells for each speaker gives a measure of robustness, in terms of how many cues are used to encode focus contrasts. Whereas speakers F1, F2 and M1 use all five explored cues, speaker F3 only uses three, and speaker M2 only uses two (i.e. duration of target word and number of prenuclear accents). The number of dark grey cells for each speaker can be seen as a measure of partitioning – that is, whether the phonetic space of the cue is partitioned into multiple regions (each corresponding to a category). For example, the duration of the target word is significantly different in the three focus conditions for speakers F3 and M1, but only allows for a single contrast in productions from speakers F1 and F2 (differentiating cases of Contrastive focus from cases of Broad or Narrow focus) and from speaker M2 (for whom duration rather differentiates cases of Broad focus from cases of Narrow or Contrastive focus).

Interestingly, Table 1 shows that the contrast between focus types is encoded by all speakers, albeit with different degrees of robustness and partitioning. In productions from speaker M2, for example, Broad focus can be distinguished from Narrow and Contrastive focus through a single cue (viz. the acoustic duration of the target word), and Contrastive focus can be distinguished from Broad and Narrow focus through a single other cue (viz. the number of prenuclear accents). While this means that the three intended categories can still be reliably decoded through their acoustic exponents, it is clear that this speaker encodes the three-way contrast in a suboptimal way – especially if compared with productions from speaker F2,

quantitative analysis, including the direction of the reported effects and results for the background condition, see Grice et al. (ms).

who differentially encodes categories using all cues (high robustness), two of which (peak alignment and height) actually allow for three-way contrasts (high partitioning).

Table 1: Encoding of contrasts between three focus categories (Broad, Narrow and Contrastive), split by cues (rows) and speakers (columns). The tilde indicates absence of statistically significant differences between focus categories.

Cue \ Speaker	F1	F2	F3	M1	M2
Peak alignment	B N~C	B N C	B N C	B N~C	B~N~C
Peak height	B N~C	B N C	B N C	B N~C	B~N~C
Duration of target word	B~N C	B~N C	B N C	B N C	B N~C
Duration of first word	B N~C	B N~C	B~N~C	B N~C	B~N~C
Number of prenuclear accents	B N~C	B~N C	B~N~C	B N~C	B~N C

The articulatory analysis provides comparable results. Figure 5 shows averaged lip aperture trajectories broken by speaker (columns), target words (rows) and focus conditions (line types). Again, trajectories are clearly distinguishable for speaker F1, for all four focus conditions, in all target words. This is not the case for productions from speaker F3, for whom only one out of four focus conditions (viz. contrastive) seems to follow a different pattern, and only for two out of three target words (Bahber and Bieber).

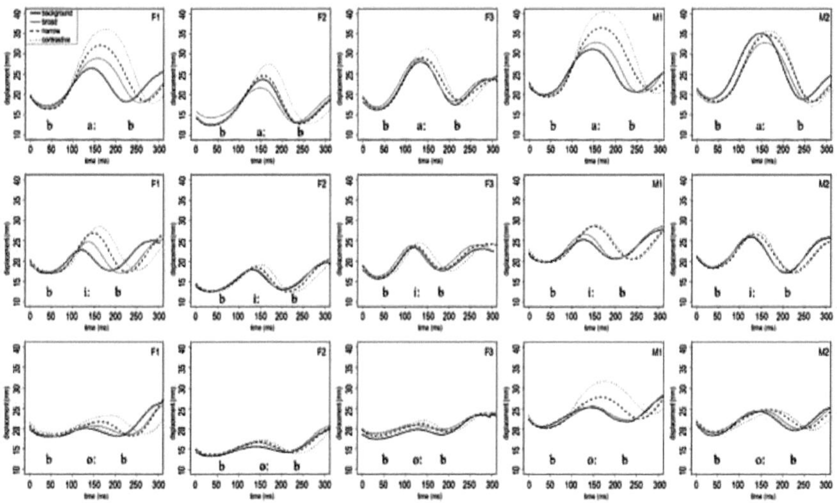

Figure 5: Averaged trajectories of lip aperture for the target words B/a:/ber, B/i:/ber *and* B/o:/ber, *separately for each speaker (F1, F2, F3, M1, M2) with different focus structures. All trajectories are aligned with the acoustic beginning of the target word. Figure 5. Reprinted from* Journal of Phonetics *44, 47–61. Mücke, D. and Grice, M., The effect of focus marking on supra-laryngeal articulation – is it mediated by accentuation? Reproduced with permission from Academic Press.*

3.2. Perception

Results from the perception task are presented as percentages of listeners' correct responses, evaluated with respect to the intended categories produced by the five speakers above (chance level: 25%). Figure 6 shows responses pooled across listeners and split by speakers. The trends are consistent with the expectations stemming from the production study. For instance, productions from speaker F2, who encoded focus robustly and distinctively, are correctly identified more often (69.32%) than productions from speaker M2 (63.7%), which had suboptimal encoding of focus (cf. 3.1.). Speakers can thus be arranged along a continuum of contrast maximization in encoding focus structures. This result is not incompatible with the notion of a "universal phonetic donor", that is, of a speaker being generally more accurate in encoding phonological contrasts, which in turn makes such contrasts easier to decode for all listeners (cf. 1.2.).

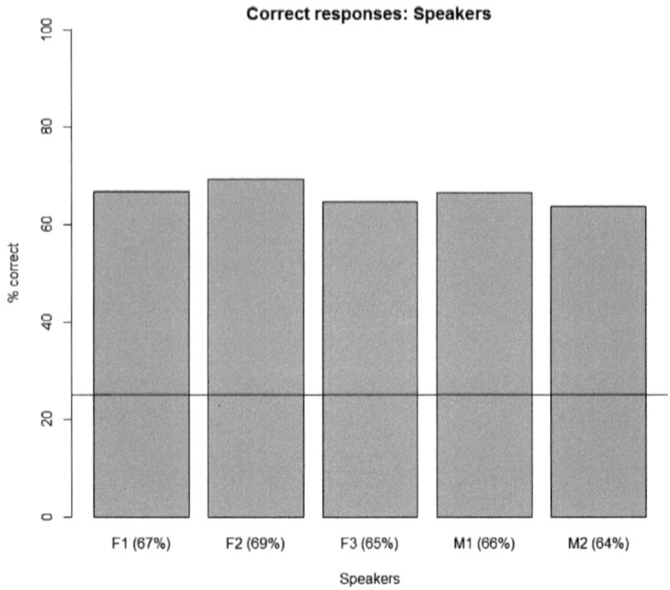

Figure 6: Percentage of correct responses from all listeners to stimuli produced by individual speakers. The horizontal line indicates chance level.

When split by individual listeners (Figure 7), results from responses to productions from all speakers indicate an even greater variability in how proficient individuals are at decoding focus structures, with one listener providing correct answers in three out of four cases (viz. BB, 74.87% correct) and another listener in just over half of the cases (viz. KS2, 55.55% correct). This is, again, compatible with the notion of a "universal phonetic recipient", that is a listener who is overall more reliable at decoding intended categories as produced by any speaker.

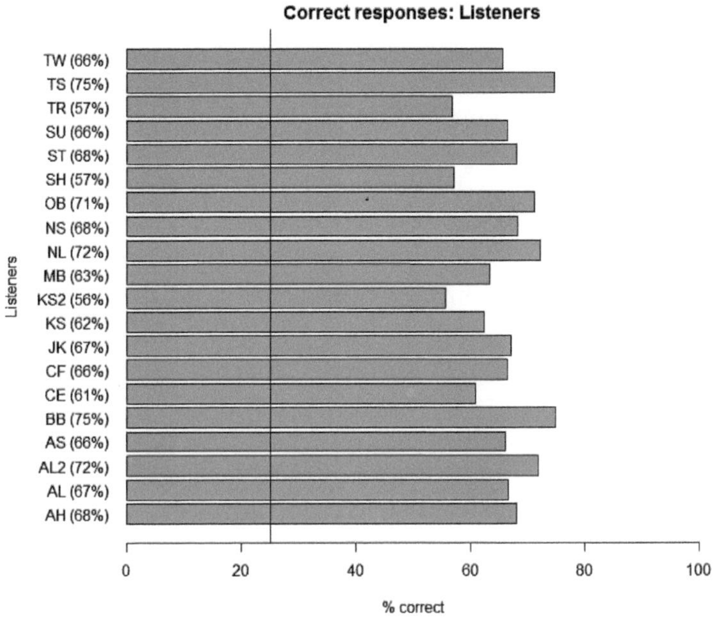

Figure 7: Percentage of correct responses to stimuli from all speakers for individual listeners. The vertical line indicates chance level.

3.3. Interaction

The qualitative analysis of the results from the perception study is thus compatible with the notion that some speakers (listeners) are overall more proficient in encoding (decoding) focus structures. In the following, we provide a quantitative evaluation of the hypothesis that some particular listeners might be particularly proficient at decoding structures as encoded by some particular speakers – that is, the hypothesis of an interaction between speaker- and listener-specific behaviour (cf. Figure 2, bottom panel).

The heat map in Figure 8 shows correct responses pooled across focus conditions and split by speakers (x-axis) and listeners (y-axis), with darker shades of grey corresponding to higher correct response scores. Average scores pooled across listeners and speakers are put in parentheses after each speaker and listener identifier on the axes, thus incorporating information from Figures 6 and 7. If any single speaker had been more intelligible to all listeners overall, we would expect one single column in the figure to be darker than the others. Similarly, had any single listener been more successful at decoding contrasts produced from all speakers, we would expect the presence of continuous darker rows in the figure. An informal analysis of the figure, however, shows that this is not the case. It is true that some columns might seem overall darker than others, thus indicating that a given speaker is more intelligible than another (e.g. F2 vs. M2), as confirmed by the average scores on the x-axis and by Figure 6. It is also true that some rows seem overall darker than others, thus indicating that a given listener is more successful than another (e.g. BB or TS vs. KS2 or SH or TR), as confirmed by the average scores on the y-axis and by Figure 7.

But Figure 8 also shows a more interesting pattern of results: The same speaker can produce contrasts which are well decoded by a certain listener, but poorly decoded by another listener. Productions from speaker F1, for example, are decoded very reliably by listeners BB and ST, but very poorly by listeners SH and CE. Similarly, the same listener can reliably decode contrasts as produced by a given speaker, while being less reliable with productions from a different speaker. Listener MB is for example very reliable when decoding contrasts produced by speaker F2 but is less reliable with productions from speaker F1.

Crucially, the same speakers and listeners can be involved in diametrically opposed patterns of results: whereas listeners CE and AL seem to over-perform on productions from speakers M1 and underperform on productions from speaker F1, listeners ST and JK seem to do the opposite (over-performing on F1 and under-performing on M1). An informal analysis of Figure 8 is thus consistent with the hypothesis of an interaction between speaker- and listener-specific behaviour.

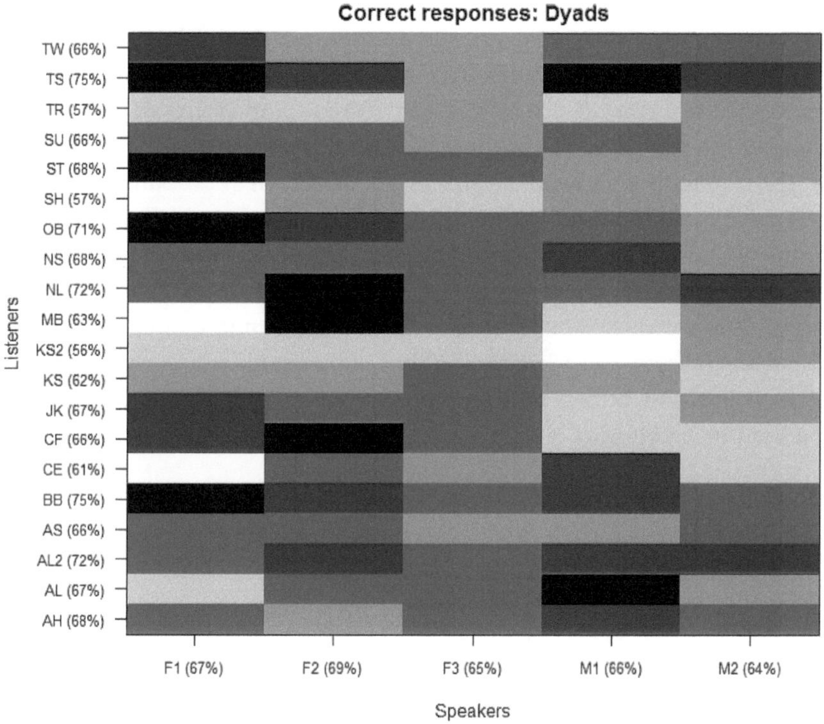

Figure 8: Percentage of correct responses for individual speakers from individual listeners.

The most straightforward way to test this hypothesis is to conceptualize such interaction as an interaction in the statistical sense. We thus used logit modelling to predict correct identification scores (on data from all focus categories) using the factors SPEAKER (from 1 to 5), LISTENER (from 1 to 20), and their interaction. This full model was compared to a null model which dropped the interaction between the two factors. A Likelihood Ratio Test yielded highly significant results ($\chi^2(76) = 145.46$, $p = 0.000003$).

Significant results were achieved also through testing based on mixed effect models. The null model included random intercepts for SPEAKER (from 1 to 5) and LISTENER (from 1 to 20) only. The full model also included random intercepts for DYAD (from 1 to 100), that is the individual pairings of speakers and listeners (e.g. speaker F1 with listener AH, speaker F1 with listener AL, et cetera). A Likelihood Ratio Test revealed a significant difference between the two models ($\chi^2(1) = 18.884$, $p = 0.00002$).

In order to quantify the dyadic interaction effects illustrated through a grey scale in Figure 8, we ranked the 100 random intercepts for DYAD, assigning the first place to the most beneficial interaction (which indicates that the listener in that dyad is particularly proficient at decoding contrasts as encoded by the speaker in that dyad) and the last place to the most detrimental. We conservatively focussed on the 10 most detrimental interactions (with DYAD random intercepts below –0.2, rankings from 91 to 100) and on the 10 most beneficial interactions (with DYAD random intercepts above 0.2, rankings from 10 to 1) only. Table 2 shows rankings, coefficients and dyads (relevant identifiers with speakers in boldface and listeners in italics, separated by a colon).

Table 2: Random intercepts for Dyads.

most detrimental ←

Rank	100	99	98	97	96	95	94	93	92	91	90	89
Coef.	–0.39	–0.33	–0.28	–0.22	–0.21	–0.21	–0.21	–0.2	–0.17	–0.17	...	...
Dyad	F1:CE	F1:MB	**F1**:*AL*	M1:CF	F1:SH	M1:KS2	**M1**:*JK*	M2:CF	M2:OB	F3:SH	...	...

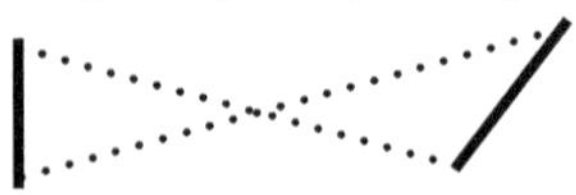

Dyad	...	...	**F1**:*JK*	M1:CE	F1:TS	**M1**:*AL*	F2:CF	F1:BB	F1:OB	F2:NL	F2:MB	F1:ST
Coef.	...	...	0.2	0.23	0.25	0.25	0.26	0.28	0.3	0.3	0.3	0.31
Rank	12	11	10	9	8	7	6	5	4	3	2	1

→ **most beneficial**

Interestingly, we found for example that listener AL is remarkably reliable at decoding productions of speaker M1 (ranking at 7) but also unreliable at decoding productions by speaker F1 (rank 98), whereas listener JK has the opposite behaviour, being particularly reliable with productions from speaker F1 (rank 10) and unreliable with M1 (rank 94). The solid lines in Figure 9 show that the same speaker (i. e. F1 or M1) can be involved in both very beneficial and very detrimental interactions, depending on the listener. The dotted lines show that same pattern for listeners (e.g. JK or AL) with respect to speakers[2].

4. Discussion

4.1. Summary of findings

Our results on the encoding and decoding of focus structures in German show the existence of *interacting speaker- and listener-specific strategies*. Specifically,

(i) There is variation among speakers with respect to how phonetic cues are used to encode focus structures, both in terms of robustness (i.e. how many cues are employed) and partitioning (i.e. how many contrasts are expressed by a single cue); cf. 3.1., Table 1.
(ii) Such variation makes the productions of some speakers more intelligible overall than the productions of other speakers; cf. 3.2., Figure 6.
(iii) Listeners vary with respect to how reliable they are in correctly identifying focus structures as intended by speakers; cf. 3.2., Figure 7.
(iv) On top of the overall trends in (ii) and (iii), we document an interaction between individual-specific strategies in production and perception. The same speaker can be more intelligible than average for one particular listener and less intelligible than average for another particular listener; cf. 3.3., Figure 8.

Since subjects do not serve as both speakers and listeners in our dataset, we could not directly verify the hypothesis of a link between accuracy in production and precision in perception, as tested for example by Perkell et al. (2004a,b). However, the result (iv) above seems to question the

2 As stated above (2.2.), subjects participating to the perception task reported normal hearing. A thorough exploration of listener-specific patterns would require ruling out hearing problems through audiometric tests. This was not possible for this study, since the materials used here were collected for independent studies, and subjects were not available for further testing. We could however perform a full audiometric test (using an Amplaid 200 audiometer) on a single listener involved in one of the crucial interactions discussed above (listener JK), and observed normal hearing for all frequency bands (i.e., no hearing loss above the 15 dB threshold).

possibility of understanding accuracy (of production and perception) in absolute terms that is observing speakers or listeners individually, outside their dyadic interactions.

4.2. Implications for linguistic theory

Apart from being relevant to intonation research, in which the interaction between speaker- and listener-specific behaviour is scarcely documented, we believe the findings above to be of interest to linguistic theory in general.

First, our findings provide additional evidence supporting claims of a complex relationship between phonetic exponents and phonological contrasts. Multiple cues are involved in the signalling of phonological categories, not only in the segmental domain (e.g. Lisker, 1986; Coleman, 2003) but also for intonational contrasts (see also Cangemi and Grice, to appear). Certain cues, such as voice onset time for voicing contrasts, or peak alignment for pitch accent type contrasts, might be particularly important in both production and perception. However, since they are weighted with respect to other (potentially underexplored) cues, they cannot be treated as the sole exponents of phonological contrasts.

Crucially, the weights associated with phonetic cues in the encoding (and decoding) of contrasts can differ across speakers (and listeners). Even domains in which individual specificity in cue weighting is largely underexplored as is the case in intonation research, are starting to acknowledge the possibility that (groups of) speakers might encode a phonological contrast by relying more or less strongly on different cues. The study by Niebuhr et al. (2011) mentioned above (1.1.) provided initial evidence in this sense, by showing that speakers of Standard Northern German express the contrast between H+L* and H* by primarily varying either peak alignment or peak shape.

Third, acknowledging the interaction between speaker- and listener-specific behaviours leads to a refined understanding of intelligibility (and proficiency) in the encoding (and decoding) of contrasts. Our findings provide evidence that, on top of overall average individual skills as encoders, the intelligibility of individual speakers is modulated by the specificities of the individuals acting as decoders. Likewise, the performance of individual listeners is affected by the specificities of the individual speakers. Our results thus call into question the metaphor "universal donors and recipients" in speech.

Moreover, the results presented here point to the necessity of exploring the cascading of individual specificity in production and perception – something along the lines of the empedoclean gnoseological principle of "like is known by like". The specific hypothesis to test would be whether an individual X, who *as a listener* has an advantage in decoding a given contrast as produced by speaker Y (rather than by speaker Z), also happens to encode the same contrast *as a speaker* by weighting cues as Y does (rather than as Z does).[3]

Finally, and perhaps most importantly, our findings strengthen the case for the impossibility of conceptualizing phonological categories in the monothetic sense. Rather than singly necessary and jointly sufficient features for category membership, phonetic cues are better understood as dimensions along which phonological categories cluster, in an individual-specific network of phonological knowledge.

Acknowledgements

The work of the first author is funded by the German Research Foundation Excellence Initiative at the University of Cologne (Emerging Group: Dynamic Structuring in Language and Communication). We would like to thank two anonymous reviewers for their useful comments and Roger Mundry (Max Planck Institute for Evolutionary Anthropology, Leipzig) for support with the statistical analysis of speaker-listener interaction. Earlier versions of this study received useful feedback at the 4th Summer School on Speech Production and Perception – Speaker-specific behaviour (Aix-en-Provence, October 2013) and at the 14th Conference on Laboratory Phonology (Tachikawa, July 2014).

References

Allen J.S., Miller, J.L., and DeSteno, D. (2003). Individual talker differences in voice-onset-time. *The Journal of the Acoustical Societcy of America,* 113(1), 544–552.

3 Research on this topic is still in its infancy; Idemaru et al. (2012, Exp. 4) failed to provide positive evidence for a within-speaker correlation between production and perception cue weights for stop length contrasts in Japanese.

Andreeva, B., and Barry, W. (2007). Cross-language and individual differences in the production and perception of syllabic prominence. *3rd annual meeting of the DFG-Priority Programme 1234*, Cologne, Germany.

Baayen, R.H. (2008). *Analyzing linguistic data. A practical introduction to statistics using R*. Cambridge: Cambridge University Press.

Boersma, P., and Weenink, D. (2010). *Praat: doing phonetics by computer.* Computer program.

Bradlow, A., Torretta, G., and Pisoni, D. (1996). Intelligibility of normal speech: I. Global and fine-grained acoustic-phonetic talker characteristics. *Speech Communication, 20*, 255–272.

Byrd, D. (2000). Articulatory vowel lengthening and coordination at phrasal junctures. *Phonetica, 57*, 3–16.

Cangemi, F., and Grice, M. (to appear). A distributional approach to categoriality in intonation transcription. Laboratory Phonology.

Cassidy, S., and Harrington, J. (2001). Multi-level annotation in the EMU speech database management system. *Speech Communication, 33*, 611–677.

Coleman, J. (2003). Discovering the acoustical correlates of phonological contrasts. *Journal of Phonetics, 31*, 351–372.

Grice, M., Baumann, S., and Benzmüller, R. (2005). German intonation in autosegmental–metrical phonology. In: Jun, S. (ed.), *Prosodic typology: The phonology of intonation and phrasing*, pp. 55–83. Oxford: Oxford University Press.

Grice, M., Ritter, S., Niemann, H., and Roettger, T. (ms). *Continuous measures provide insights into the nature of intonation as a marker of focus type*. Unpublished manuscript, University of Cologne, Germany.

Hazan, V., and Baker, R. (2011). Is consonant perception linked to within-category dispersion or across-category distance? In *Proceedings of the 17th International Congress of Phonetic Sciences*, Hong Kong, 839–842.

Hazan, V., Romeo, R., and Pettinato, M. (2013). The impact of variation in phoneme category structure on consonant intelligibility. In *Proceedings ICA Montreal, 19*, 1–6.

Idemaru, K., Holt, L., and Seltman, H. (2012). Individual differences in cue weights are stable across time: The case of Japanese stop lengths. *The Journal of the Acoustical Socіecty of America, 132*(6), 3950–3964.

Jun, S. (ed.) (2014). *Prosodic typology II: The phonology of intonation and phrasing*. Oxford: Oxford University Press.

Lakoff, G. (1987). *Women, fire and dangerous things: What categories reveal about the mind.* Chicago and London: The University of Chicago Press.

Lisker, L., and Abramson, A. (1964). A cross-language study of voicing in initial stops: Acoustical measurements. *Word, 20*(3), 527–565.

Lisker, L. (1986). "Voicing" in English: A catalogue of acoustic features signaling /b/ versus /p/ in trochees. *Language and Speech, 29*(1), 3–11.

Mücke, D., and Grice, M. (2014). The effect of focus marking on supra-laryngeal articulation – is it mediated by accentuation? *Journal of Phonetics, 44*, 47–61.

Newman, R., Clouse, S., and Burnham, J. (2001). Perceptual consequences of within-talker variability in fricative production. *The Journal of the Acoustical Society of America, 109*, 1181–1196.

Niebuhr, O., D'Imperio, M., Gili Fivela, B., and Cangemi, F. (2011). Are there "shapers" and "aligners"? Individual differences in signaling pitch accent category. In *Proceedings of the 17th International Congress of Phonetic Sciences*, Hong Kong, 120–123.

Perception Research Systems (2007). *Paradigm Stimulus Presentation.* Computer program.

Perkell, J., Guenther, F., Lane, H., Matthies, M., Stockmann, E., Tiede, M., and Zandipour, M. (2004a). The distinctness of speakers' productions of vowel contrasts is related to their discrimination of the contrasts. *The Journal of the Acoustical Society of America, 116*, 2338–2344.

Perkell, J., Matthies, M., Tiede, M., Lane, H., Zandipour, M., Marrone, N., Stockmann, E., and Guenther, F. (2004b). The distinctness of speakers' /s/-/ʃ/ contrast is related to their auditory discrimination and use of an articulatory saturation effect. *Journal of Speech, Language and Hearing Research, 47*, 1259–1269.

Schultz, A., Francis, A., and Llanos, F. (2012). Differential cue weighting in perception and production of consonant voicing. *The Journal of the Acoustical Society of America, 132*(2), EL95–101.

Iris Chuoying Ouyang and Elsi Kaiser
University of Southern California Los Angeles

Individual Differences in the Prosodic Encoding of Informativity

Abstract: This chapter presents a psycholinguistic production study that investigates individual differences in the prosodic encoding of informativity. In particular, it examines how the shapes of f0 contours and the sizes/ranges of f0 excursions are influenced by the interaction between information structure and information-theoretic properties. We focus on two types of information structure, namely new-information narrow focus and corrective narrow focus, and two kinds of information-theoretic properties, namely word frequency and contextual probability. We analyze (i) group trends, (ii) between-subject variability as well as (iii) within-subject variability, and thereby identify speaker-specific effects. Our results show that word frequency and contextual probability modulate the f0 movement associated with new-information narrow focus and corrective narrow focus respectively (see also Ouyang and Kaiser, 2014). Furthermore, f0 ranges appear to be more informative than f0 shapes in reflecting informativity across speakers. Specifically, speakers seem to have individual 'preferences' regarding f0 shapes, the f0 ranges they use for an utterance, and the magnitude of differences in f0 ranges by which they mark information-structural distinctions. In contrast, there is more universality over the *directions of differences in f0 ranges between information-structural types*. Our findings highlight the importance of disentangling information structure and information-theoretic factors and examining both inter- and intra-speaker variability.

1. Introduction

It is widely accepted that prosody can reflect the extent to which a linguistic element is 'informative'. Prior work has approached the relationship between prosody and informativity from various angles, of which two popular ones are **information structure** (e.g. Breen et al., 2010; Brown, 1983; Cooper et al., 1985; Couper-Kuhlen, 1984; Eady and Cooper, 1986; Hay et al., 2006; Katz and Selkirk, 2011; Krahmer and Swerts, 2001; Ladd, 1996; Pierrehumbert and Hirschberg, 1990) and **information theory** (e.g. Aylett and Turk, 2004; Baker and Bradlow, 2009; Bell et al., 2003; Bell et al., 2009; Calhoun, 2010; Clopper and Pierrehumbert, 2008; Gregory et al., 1999; Lieberman, 1963; Munson and Soloman, 2004; Pan and Hirschberg, 2000; Pitrelli, 2004; Pluymaekers et al., 2005a, 2005b; Scarborough, 2010; van Son et al., 1998; Wright, 2004). It has been found that the acoustic properties of an utterance such as duration, f0, intensity, and spectral characteristics provide cues for the relative informativity of its components (see Wagner and Watson, 2010 for a review). However, existing studies have also noted that speakers differ in their acoustic characteristics and the prosodic patterns they use to signal linguistic categories (e. g. e.g., Allen et al., 2003; Dahan and Bernard, 1996; Ferguson, 2004; Ferguson and Kewley-Port, 2007; Loakes and McDougall, 2010; Niebuhr et al., 2011; Smith and Hawkins, 2012; Theodore et al., 2007; Trouvain and Grice, 1999). In this section, we will first discuss the previous research on prosody from the perspectives of information structure, information theory, and individual differences. Then, we will describe the aims and predictions of our study, which integrates the insights from these different traditions of research and furthers our understanding of prosody and informativity.

1.1. Prosodic prominence and information structure

In the information-structure-based tradition, acoustic prominence is associated with linguistic material in the foreground, or in focus – broadly speaking, material that adds new information to the conversation. Depending on the preceding discourse, speakers may emphasize particular words in an utterance to direct their addressee's attention to the important message they are trying to convey. It has been found that some types of information structure differ acoustically from each other. For instance, consider the word 'toys' in the following contexts:

(1) a. What did David find on the stairs?
    b. He found toys on the stairs. ['toys' = narrow, new-information focus]

(2) a. Did David find toys on the stairs?
    b. Yes, he found toys on the stairs. ['toys' = given information, unfocused]

(3) a. What happened?
b. David found toys on the stairs. ['toys' = wide, new-information focus]

In response to (1a), 'toys' in (1b) is in new-information focus, as it conveys information that has not been mentioned and cannot be inferred from the preceding discourse. In contrast, the same word 'toys' in (2b) responding to (2a) is unfocused, given information, because what it conveys has been expressed in the preceding discourse (e.g. Prince, 1992; Rooth, 1992). Furthermore, 'toys' in (1b) is narrowly focused new information, since it is the only component of the utterance that introduces new information to the conversation. However, the same word 'toys' in (3b) in response to (3a) is new information in wide focus, because the entire utterance with multiple components including toys is in new-information focus (e.g. Gussenhoven, 1983; Selkirk, 1984). It has been shown that new elements are acoustically more prominent than given elements (e.g. Brown, 1983; Eady and Cooper, 1986; Hay et al., 2006; Krahmer and Swerts, 2001; Ladd, 1996), and that material in narrow new-information focus is acoustically more prominent than the same material in wide new-information focus (e.g. Breen et al., 2010; Eady and Cooper, 1986).

Another kind of information structure that has been extensively studied is contrastive focus, of which various subtypes have been identified (e.g. Vallduví and Vilkuna, 1998). For example, two common types of contrastive focus, both involving explicit alternatives in the preceding discourse, are shown in (4–5). 'Toys' in (4b) responding to (4a) picks 'toys' from the set consisting of 'books' and 'toys' that has been established via (4a), and 'toys' in (5b) responding to (5a) is intended to contradicts the information 'socks' that has been conveyed via (5a). In this study, we concentrate on the latter type of contrastive focus, which has been referred to as corrective focus (e.g. Dik, 1997). We chose this subtype because its information-structural properties are well-understood and it is prevalent in communication. Contrastive/corrective elements have been shown to receive greater acoustic prominence than non-contrastive/non-corrective elements, whether they are given or new material in the discourse (e.g. Breen et al., 2010; Cooper et al., 1985; Couper-Kuhlen, 1984; Katz and Selkirk, 2011; Krahmer and Swerts, 2001).

(4) a. Did David find books or toys on the stairs?
b. He found toys on the stairs. [‘toys’ = narrow, contrastive focus]

(5) a. Did David find socks on the stairs?
b. No, he found toys on the stairs. [‘toys’ = narrow, contrastive/ corrective focus]

Various acoustic properties have been found to reflect information-structural salience, including types or presence of accents on and after the focused element (e.g. Krahmer and Swerts, 2001; Ladd, 1996; Pierrehumbert and Hirschberg, 1990), expanded vowel space and increased formant movement in the focused element (e.g. Hay et al., 2006), increased duration, f0[1], intensity, and more f0 protrusion during the focused element, more f0 compression following the focused element (e.g. Breen et al., 2010; Brown, 1983; Couper-Kuhlen, 1984; Katz and Selkirk, 2011), decreased duration, f0 and intensity preceding the focused element (e.g. Eady and Cooper, 1986), and a sudden drop or sharper fall within or following the focused element (e.g. Cooper et al., 1985; Couper-Kuhlen, 1984; Eady and Cooper, 1986).

1.2. Prosodic prominence and information-theoretic factors

In addition to work from the information-structural perspective, there is also research in the information-theoretic tradition, where a correlation has been found between acoustic reduction and the redundancy, or the predictability of a linguistic element. Depending on what is more (or less) common in the language or the given linguistic environment, certain elements may be pronounced with more or less acoustic prominence. A wide variety of probabilistic measurements have been used to represent the predictability of a segment, phoneme, or word. Examples include context-independent

1 In Breen et al. (2010), focus breadth (narrow vs. wide) and contrastiveness (corrective vs. non-corrective) have opposite effects on f0. Narrow focus is marked with *higher* mean and maximum f0 than wide focus, while correctively focused word is produced with *lower* mean and maximum f0 than non-correctively focused word. This finding about contrastiveness diverges from other previous research.

properties such as frequency and neighborhood density (e.g. Munson and Soloman, 2004; Pitrelli, 2004; Scarborough, 2010; Wright, 2004) and context-dependent properties such as joint probability, conditional probability, mutual information, and semantic predictability (e.g. Bell et al., 2003; Clopper and Pierrehumbert, 2008; Lieberman, 1963; Pan and Hirschberg, 2000; Scarborough, 2010; van Son et al., 1998). Elements that occur more frequently or have more neighbors (i.e. items that are similar to each other due to overlapping features) in the language are acoustically more reduced than elements that occur less frequently or have fewer neighbors. Likewise, elements that are more likely to occur in a particular environment (based on adjacent items or semantic context) receive larger acoustic reduction than elements that are less likely to occur in the environment. Research has found information-theoretic predictability being realized with decreased duration and amplitude (e.g. Bell et al., 2003; Lieberman, 1963), lower likelihood of accentuation (e.g. Pan and Hirschberg, 2000; Pitrelli, 2004), lower center of gravity of the power spectrum (CoG), less extreme distance between the first and second formants (e.g. van Son et al., 1998), shorter vowels, and less dispersed vowel space (e.g. Clopper and Pierrehumbert, 2008; Munson and Soloman, 2004; Scarborough, 2010; Wright, 2004).

1.3. Connections between information-structural and information-theoretic approaches

While the information-structural and the information-theoretic traditions focus on different factors of informativity from distinct perspectives, they have found similar prosodic patterns that signal the relative degree of informativity between linguistic elements (see sections 1.1. and 1.2.). A higher degree of informativity in general results in more exhaustive use of a prosodic space, whichever acoustic dimension it is that a particular study examines. This leads us to the question of how information structure and information-theoretic properties interact in influencing prosody: Do they simply have additive effects, or do they interact in a non-additive way? To our knowledge, only a limited number of studies have investigated both of these two types of informativity (e.g., Aylett and Turk, 2004; Baker and Bradlow, 2009; Bell et al., 2009; Calhoun, 2010; Gregory et al., 1999; Pluymaekers et al., 2005a, 2005b). Most of these studies take an information-theoretic approach that includes the repeated use of words as a redundancy factor. Repeated words are by definition given, or at least not entirely new, information, and thus the information-theoretic notion of repetition can be regarded as givenness in an information-structural view (e.g. Fowler and Housum, 1987). The effect of word repetition, over and above (other) information-theoretic factors, has been found on different kinds of linguistic units. Aylett and Turk (2004) measure how many times a referent has been previously mentioned, and show that syllable duration decreases as the order of mention increases, in addition to the effects of word frequency and syllable conditional trigram probability. For suffixed words in Dutch, Pluymaekers, Ernestus, and Baayen (2005b) measure how many times a word has been uttered, and show that repetition significantly reduces the duration of suffixes and marginally reduces the duration of stems and entire words, in addition to the effects of mutual information with the adjacent words. Bell, Brenier, Gregory, Girand and Jurafsky (2009) find that, in English, content words are shorter when repeated, more frequent, or more predictable from the following word, while function words are not so affected by repetition and word frequency, but are affected by the predictability from the following word. The predictability from the preceding word only shortens very frequent function words. Lastly, Gregory, Raymond, Bell, Fosler-Lussier and Jurafsky (1999) find that word duration decreases as the following redundancy factors increase: word frequency, mutual information, conditional bigram probability, semantic relatedness, and repetition. In sum, word repetition has been shown to cause shortening at the syllable, morpheme and word levels, even when we take into account word frequency and other statistical-probabilistic factors based on adjacent items or semantic context.

To the best of our knowledge, there is only one existing study that addresses the *interaction* between word repetition and (other) information-theoretic factors. In a production experiment where participants read a number of paragraphs twice, Baker and Bradlow (2009) find that word frequency influences the amount of reduction a word undergoes when it is mentioned for the second time. Higher-frequency words exhibit more shortening upon second mention than lower-frequency words, when word length is controlled. Furthermore, this interaction is only found in plain speech, i.e., when participants are instructed to speak as if they are talking to someone familiar with their voice and speech patterns. It does not occur in clear speech, i.e., when participants are instructed to speak as if they are talking to a listener with a hearing loss or to a non-native speaker learning their language. From the perspective of information structure, this finding can be restated as: the duration cue for new information (i.e. first mention) is weaker in lower-frequency words, and weaker in clear speech compared to plain speech. Thus, there seems to be a saturation effect such that the prosodic cues for information structure are weakened when information-theoretic factors also demand prosodic prominence. However, it remains unclear whether other kinds of information-theoretic factors, such as contextual probability, have a similar impact and whether other kinds of information structure, such as corrective focus, are affected in a similar way. Calhoun (2010) shows that whether a word carries a nuclear accent, non-nuclear accent, or no accent can be predicted using models including word frequency, bigram probability, the presence/absence of focus, as well as other factors. Nevertheless, no interaction between these factors is mentioned. In sum, it is not yet well-understood how information-theoretic properties and information structure interact to influence prosody.

1.4. Individual differences in prosody and the prosodic encoding of informativity

In addition to the interaction between different types of informativity, another important factor that influences an utterance's prosodic representation is individual differences. Research has shown that speakers should not be assumed to be homogenous even though they speak the same language. Speakers can differ in their ways of marking the linguistic distinction in question using duration, f0, intensity and spectral parameters. To name a few, individual differences have been investigated in the duration and spectral cues for word boundary (e.g. Smith and Hawkins, 2012), in voice-onset-time (VOT) for stop consonants (e.g. Allen et al., 2003; Loakes and McDougall, 2010), and how VOT is affected by other factors such as speech rate and place of articulation (e.g. Theodore et al., 2007).

It appears that between-subject variability can occur qualitatively and quantitatively, both on a general level and in specific cases. Along a given acoustic dimension, participants have different ranges of absolute values, produce different sizes and directions of variation between and within linguistic categories, and use different kinds and numbers of strategies to signal a linguistic contrast. For example, in a study where participants were asked to speak at self-selected fast, normal and slow rates, some people's fast rates were similar to some others' slow rates in terms of the number of syllables they produced per second. Moreover, the participants differed in how they altered their speech rate: while some people produced more syllables a second for a faster rate, some others produced longer pauses for a slower rate (Trouvain and Grice, 1999 for German). In a study by Dahan and Bernard (1996) on French emphatic accent with four participants, some people increased f0 to a greater extent than others. The participants also differed in where and how they used intensity to signal emphasis. For the emphasized element in a sentence, one person increased the intensity, another person decreased it, and two other people produced no difference. In the sentence region preceding the emphasized element, three people decreased the intensity, while one person produced no differences. Lastly, everyone decreased the intensity in the sentence region following the emphasized element (Dahan and Bernard, 1996).

In addition to individual differences in the modulation of duration, pauses, f0 and intensity, work by Niebuhr et al. (2011) found evidence for individual differences on the realization of pitch accent categories in Standard Northern German (H* vs. H+L*), Neapolitan Italian (L+H* vs. L*+H) and Pisa Italian (H* vs. H*+L). They also found that Standard Northern German and Neapolitan Italian speakers used different strategies in terms of the alignment and shapes of f0 contours: some people produced systematic differences in the location of the f0 peak with respect to the target syllable, while others produced systematic differences in how steep and large the f0 rise or fall was. In contrast, Pisa Italian speakers only differed in cue strength: those who made greater alignment differences also made greater differences in shapes.

Individual differences also exist in the strategies people use for increasing the audibility/intelligibility of their speech. In a study where participants were first asked to speak normally and then asked to speak as they would if they were talking to a hearing-impaired person, individual differences were observed. According to normal-hearing listeners in a perception study, some of the speakers significantly improved their vowel intelligibility while others did not. It turns out that the former group of speakers increased their vowel duration and raised their F2 for front vowels to a greater extent than the latter group. Also, the former group expanded their vowel space in the F1 dimension, while the latter group did not (Ferguson, 2004; Ferguson and Kewley-Port, 2007). In sum, empirical evidence suggests that speakers may differ from one another substantially in terms of whether and how particular acoustic markers correlate with particular linguistic factors.

In addition to the studies that explicitly focus on individual differences, research whose primary focus is not on individual differences has also led to observations about between-subject variability, i. e. how individuals differ. For example, it has been noted that participants differed in their duration and spectral cues for the edges of prosodic domains (e. g. Fougeron and Keating, 1997; Krivokapić and Byrd, 2012; Korean: Cho and Keating, 2001), in their pausing and lengthening cues for levels of discourse structure (e. g. word vs. clause vs. paragraph in Dutch, see van Donzel and Beinum, 1996), and in the effect of word prosodic structure on vowel duration (e. g. Rietveld et al., 2004 for Dutch).

More specifically related to informativity, Krahmer and Swerts (2001) investigated the intonational cues for the distinctions between contrastive focus, non-contrastive focus, and given information in Dutch. An interactive task was used, where participants worked in pairs to complete dialogues. It was found that some participants' prosodic behavior ignored their partner's contribution and instead prosodically marked elements that were contrastive to their own last utterance. These participants also tended to end their utterances with a high boundary tone (H%), which is generally interpreted as signaling the speaker's intention to hold the turn. Thereby, these 'egocentric' participants made the exceptional cases in the data.

In related work on focus types, Andreeva et al. (2007) investigated the cues in duration, f0, intensity and vowel quality for the distinctions between narrow contrastive focus, narrow non-contrastive focus, and wide focus in German. They note that some participants produced larger differences than others, and some participants also used one parameter to a greater extent than another. Thus, individual participants had their own tendencies and strategies in producing prosodic prominence. Other than these sparse observations, little is known about the extent or nature of individual differences regarding the prosodic encoding of informativity.

1.5. The present study: Aims and expected outcome

The previous research discussed in sections 1.1. to 1.3. shows that an utterance's prosodic representation depends on how informative each of its constituents is. Information-structural status, such as being in narrow focus, and information-theoretic properties, such as lexical frequency and contextual probability, both play a role in prosody. It is striking that little attention has been paid to the potential interaction between information structure and information-theoretic factors, given the considerable efforts that have been devoted to both kinds of factors separately. To shed light on this issue, we conducted a psycholinguistic production study (see Ouyang and Kaiser, 2014 for an earlier discussion of this study) to investigate whether information structure and information-theoretic factors interact in determining a word's prosodic prominence, and if so, whether different information-structural types interact with different information-theoretic factors in similar ways. For instance, could it be that the prosodic cues for new-information vs. corrective focus differ in terms of whether they are sensitive to word frequency vs. contextual probability?

Since prior work has found that the effect of givenness on duration is stronger when the repeated words are high-frequency (Baker and Bradlow, 2009), we hypothesized that the prosodic effect of information structure would be stronger in words with low informativity in the information-theoretic dimensions. In other words, the prosodic cues for information structure might be weakened when other factors – such as information-theoretical properties – also demand prosodic prominence. Building on Baker and Bradlow (2009), our study explored effects of word frequency and narrow new-information focus. We also looked at the effects of another information-theoretic factor, namely contextual probability, as well as another type of information structure, namely narrow corrective focus. Including multiple factors of each kind of informativity allowed us to investigate the potentially complex interactions among them. Specifically, we expected that narrow focus would be prosodically distinct from wide focus when the target word is highly frequent and/or highly contextually probable (i. e. has low information-theoretic informativity). In contrast, when the target word is low-frequency and/or low-probability (i. e. has high information-theoretic informativity), we predicted that the prosodic distinctions between narrow and wide focus might be weakened or even absent: prosodic reflexes of information structure might be observed in only one or perhaps in neither of the two narrow-focus conditions. If these predictions are borne out, we can then look into whether different information-structural types (i. e. corrective vs. new) could react differently to different information-theoretic factors (i. e. lexical frequency vs. contextual probability).

In addition to the general trends among speakers, the discussion in section 1.4. shows that speakers differ in their acoustic realization of prosody. As there is not a lot of prior work focusing on individual differences in sentence prosody, we first wanted to see, on a general level, whether our results fit with the previous findings that sentence prosody is susceptible to speaker-specific effects. We then also looked more closely at whether and how individual differences manifested themselves in the prosodic encoding of informativity. Roughly speaking, we expected individual differences in all aspects investigated, because existing research on other prosody-related topics (as discussed in the preceding sections) has found both qualitative and quantitative variability among the participants of a study, in terms of the range and characteristics of cues a participant produces along an acoustic dimension as well as the size and direction of acoustic differences that a participant produces to signal a linguistic contrast (Andreeva et al., 2007; Dahan and Bernard, 1996; Ferguson and Kewley-Port, 2007; Krahmer and Swerts 2001; Niebuhr et al., 2011; Trouvain and Grice, 1999). Specifically, we expected our participants to differ in whether they made distinctions between narrow and wide focus in a given information-theoretic condition, whether they increased or decreased prosodic prominence for a given region of the sentence, to what extent they vary prosodic prominence to convey the informativity of a word, as well as the overall prosody of their utterances.

In terms of the acoustic correlates of prosodic prominence, we focused on (i) the *shape of an f0 contour* and (ii) the *size of excursions in an f0 contour* (which will be called 'f0 range' henceforth). We chose f0 because it is an acoustic dimension that has been extensively studied in the information-structural tradition yet not much so in the information-theoretic tradition. In other words, by conducting this study, we also hoped to provide further evidence for the effects of information-theoretic factors on f0. Furthermore, because there are studies showing that intonational categories (e.g. H*,

L+H*) do not necessarily map straightforwardly onto focus types (e.g. Katz and Selkirk, 2011; Krahmer and Swerts, 2001, Watson et al., 2008), we did not take an intonational-phonological approach (e.g. Ladd, 1996; Pierrehumbert and Hirschberg, 1990). Based on previous research, a good indicator of narrow focus seems to be a relatively exhaustive use of the acoustic space. In the f0 dimension, as mentioned in section 1.1., it has been found that narrow focus differs from wide focus in having greater f0 protrusion or higher f0 on the narrowly focused element, greater f0 compression or sharper f0 fall following the focused element, and lower f0 preceding the focused element (Breen et al., 2010; Brown, 1983; Cooper et al., 1985; Couper-Kuhlen, 1984; Eady and Cooper, 1986; Katz and Selkirk, 2011). Therefore, we quantitatively measured both f0 shapes and f0 ranges, which presumably would capture the level of prominence in the f0 dimension.

In our study, the object of a sentence is the narrowly focused word in the discourse. Therefore, we expected narrow focus to influence prosody in the sentence region containing the object and the words immediately preceding and following it. Specifically, we predicted that the f0 movement of this sentence region would be bigger, or at least not smaller, in the narrow-focus conditions than the wide-focus condition. Also, we expected that individual participants would differ in the f0 shapes and ranges they produced in general, and the sizes and directions of differences they produced for information-structural distinctions.

2. Experiment

We conducted a production study with an interactive set-up. An earlier, abbreviated discussion of this experiment is available in Ouyang and Kaiser (2014), where we discuss some of the group results but do not explore any issues related to individual variation. Each trial consisted of a read-aloud task and a subsequent selection task. In both tasks, participants interacted with a partner, who was a lab assistant. The read-aloud task provided the critical recordings: the target sentences were produced by the participants during the read-aloud task. The selection task was included to engage both people in the read-aloud task: paying attention to what the other person said in the read-aloud task was necessary to successfully perform the selection task. (We do not discuss the selection task in detail here because it is not relevant for the results, but people essentially had to pick the correct items from a larger array).

2.1. Design and procedures

Participants worked with a partner in reading aloud sentence pairs. Each sentence pair consisted of a question spoken by the partner (Sentence A) and a response (the critical sentence) spoken by the participant (Sentence B), as shown in (1–3) below. Participants saw Sentence B on a computer screen when it was their turn to speak. The target sentences (Sentence B on target trials) are transitive clauses with the following structure: a third-person plural pronoun subject, a simple past tense verb, an object, and a prepositional phrase indicating a location. The critical word we focus on is the object of each target sentence (e.g. *balls*). The experiment had 48 target trials; each participant encountered four items in each condition and did not see any item more than once. A full list of the target sentences can be found in Appendix 1. There were also 48 filler trials. The dependent variable we measured was the f0 values of an utterance.

(1) NARROW CORRECTIVE FOCUS

A: I heard that Dawn and Alice got gloves at the sports store.

B: No, they got [balls]$_{\text{CORRECTIVE FOCUS}}$ at the sports store.

(2) NARROW NEW-INFORMATION FOCUS

A: What did Rachel and Carolyn get at the sports store?

B: They got [balls]$_{\text{NEW-INFO FOCUS}}$ at the sports store.

(3) WIDE/VP FOCUS

A: What did Angela and Joyce do?

B: They [got balls at the sports store]$_{\text{NEW-INFO FOCUS.}}$

To investigate whether information-theoretic factors interact with information structure in shaping the prosody of an utterance, we manipulated (i) the lexical frequency of the object noun, (ii) whether the object was probable in the context of the preceding verb and the following location, and (iii) the object's informational-structural status in relation to the question. Thus, a within-subject design with three independent variables was implemented: (**i**) word frequency (with two levels: high or low frequency), (**ii**) contextual probability (with two levels: high or low probability), and (**iii**) focus type (with three levels: narrow corrective focus, narrow new-information focus, or wide/VP focus).

We manipulated the *focus type* of the critical noun by means of the question asked by the partner, as shown in (1)–(3). In the wide/VP focus condition (ex. 3), the question asks about the content of the entire verb phase (i.e., *what did X do?*), and the answer spoken by the participant provides this information. Thus, the whole VP (e.g., *got balls at the sports store*) is new information. In the narrow new-information focus condition (ex. 2), in contrast, the question asks for the object of the transitive verb, and therefore only the object is new information in this condition. Finally, in the narrow corrective focus condition (ex. 1), the partner makes a statement where the object is incorrect (as signaled to the participant by the sentence on their screen), and thus the object in the participant's response is correctively focused.

The *contextual probability* of the critical words was estimated through a web-based norming study. Four verb-location contexts and eight objects were ultimately selected for the target sentences, as shown in Table 2. Each of the eight target nouns functioned as a probable object in some contexts and an improbable object in other contexts. This allows us to ensure that any effects of contextual probability cannot be attributed to idiosyncratic properties of specific nouns. Another four nouns were selected to be the 'incorrect' objects in the question that elicited corrective focus (e.g., *gloves* in ex. 1). These nouns had a contextual probability between the high-probability and low-probability critical words and a word frequency between the high-frequency and low-frequency critical words.

The *word frequency* of the object nouns was determined according to the SUBTLEXus database (Brysbaert and New, 2009). SUBTLEXus provides word frequency measures on the basis of American English subtitles, and contains 51 million words in total. The critical words in the high-frequency conditions ranged in a frequency from 67.76 to 40.16 per million, while those in the low-frequency conditions ranged in a frequency from 13.22 to 0.41 per million, as shown in Appendix 2.

Table 2: Manipulation of word frequency and contextual probability

VERB	OBJECT				LOCATION
	Frequent & Probable	Infrequent & Probable	Frequent & Improbable	Infrequent & Improbable	
got	*balls*	*cleats*	*fish*	*toys*	*at the sports store*
kicked	*cars*	*cans*	*books*	*shells*	*in the garage*
found	*fish*	*shells*	*balls*	*cans*	*in the sea*
found	*books*	*toys*	*cars*	*cleats*	*on the stairs*

2.2. Participants

Sixteen native speakers of American English participated in this study. All participants, 11 female and 5 male, were students at the University of Southern California. Two lab assistants interacted with participants in this study. Both lab assistants were female native speakers of American English and students at the University of Southern California.

2.3. Data analysis

768 utterances were collected from the 16 participants, each producing 48 target responses. Out of the full set of data, 43 utterances (5.6%) were not included in the data analysis, due to speech errors (16 utterances), disfluencies (6 utterances) and technical issues with the audio recordings (21 utterances).

F0 measurements were obtained using the STRAIGHT algorithm (Kawahara et al., 1998) through the VoiceSauce program (Shue et al., 2011). The raw f0 values were then smoothed (*smoothn* in MATLAB: Garcia, 2010) to remove f0 tracking errors and segmental effects. The smoothed values were then converted into a semitone scale, as semitones reflect pitch perception better than the Hertz scale (e.g. Nolan, 2003). Finally, the data were normalized by subject using z-scores, to factor out individual differences in f0 registers (e.g., women usually have wider and higher registers of f0 than men). The z-scores represented each data point in terms of its number of standard deviations above or below the mean across all utterances produced by a given speaker.

To investigate whether different levels of word frequency and contextual probability influence the prosodic encoding of narrow focus in different ways, we examined the effects of narrow focus in the four conditions of word frequency and contextual probability separately: high-frequency words in high-probability contexts, high-frequency words in low-probability contexts, low-frequency words in high-probability contexts, and low-frequency words in low-probability contexts[2]. As the prosodic effects of narrow focus were expected in the focused word and the words immediately before and after it (see section 1.5. for the predictions of this study), we examined these regions of a sentence. Specifically, we analyzed f0 shapes and ranges for the following three intervals: Pre-Focus (verb), Focus (object), and Post-Focus (the region from preposition to article), from Pre-Focus to Focus (verb and object), and from Focus to Post-Focus (the region from object to article)[3]. F0 ranges were calculated by subtracting the minimum f0 from the maximum f0 in each interval. Since the f0 measurements have been normalized based on a given speaker's f0 register, a larger f0 range indicates that the participant was employing a bigger proportion of his or her f0 register.

To examine the *shapes of f0 contours*, we used a smoothing spline ANOVA approach. The smoothing spline ANOVA fits regression to continuous data to test differences between curves (Gu, 2002). We plotted the best-fitted curves with 95% confidence intervals (1.96 standard errors). The best-fitted values in a regression analysis can be interpreted as representing the average patterns of the data being modeled. Two conditions can be considered as being significantly different if the 95% confidence intervals of their best-fitted values do not overlap. Similar approaches have been used for other kinds of continuous data in phonetics, such as tongue shapes (e.g. Davidson, 2006) and formants (e.g. Baker, 2006). We first extracted 20 data points with equal time spacing from each of the three consecutive

2 We did not directly compare different levels of word frequency or contextual probability, because identical sentences existed only between different types of focus. This is an intrinsic property of the design, due to the manipulation of word frequency and contextual probability.

3 We did not statistically analyze the head noun of the prepositional phrase because it was at the end of a sentence, where f0 varied considerably due to factors outside the scope of this study such as dialects (e.g. Ching, 1982) and turn transition cues (e.g. Wennerstrom and Siegel, 2003).

intervals: Pre-Focus, Focus and Post-Focus. Mixed-effects smoothing spline ANOVA models were then performed with Focus Type, Time and their Interaction as fixed effects (*gss* in R: Gu, 2014). In the analysis of group patterns (presented in section 3.), Subject and Item were included as random intercepts. For the analysis of individual patterns (presented in section 4.1.), models were performed on each participant's data separately, and Item was included as a random intercept.

For *f0 ranges*, mixed-effects models were conducted on f0 ranges (*lme4* in R: Bates et al., 2014; *lmerTest* in R: Kuznetsova et al., 2015). In the analysis of group patterns (presented in section 3.), Focus Type was included as a fixed effect, and Subject and Item were included as random effects. When specifying the structure of random effects, we started with a full model (i.e. including intercepts and slopes for Subject and Item), and if it failed to converge, we reduced the Subject slopes and/or the Item slopes until the model converged. All the group models that converged had random intercepts for Subject and Item. In the analysis of individual patterns, Subject was included as a fixed effect and Item was included as a random effect when we looked at a speaker's overall f0 ranges regardless of the condition (presented in section 4.1.). This model had both a random intercept and random slopes for Item. Finally, for the directionality and magnitude of differences in f0 ranges between conditions (presented in section 4.2.), we mainly focused on descriptive statistics, because the numbers of observations became relatively low when the data were split into small subsets by both subject and condition.

3. Group results

Overall, the predictions outlined in section 1.5. about the general trends were borne out, as can be seen in Figures 1–4, which shows the smoothing spline ANOVA results. In terms of **f0 shapes**, the three types of focus do not significantly differ in the Pre-Focus interval (the first section marked on the x-axis). Significant differences in f0 shapes start emerging towards the end of the Focus interval (the middle section marked on the x-axis) and continue for most of the Post-Focus interval (the last section marked on the x-axis). Narrow corrective focus (solid lines) and narrow new-information focus (dashed lines) have a steeper f0 drop than wide focus (dotted lines) in some cases, depending on the narrowly-focused word's frequency and contextual probability. More specifically, when the word is high-frequency and occurs in a probable context (*got balls at the sports store*), both types of narrow focus differ significantly from wide focus (Figure 1, labelled 'High Freq + High Prob'). However, when a high-frequency word is focused in an improbable context (*got fish at the sports store*), only new-information focus differs significantly from wide focus; corrective focus patterns with wide focus (Figure 2, 'High Freq + Low Prob'). In contrast, when the word is lexically infrequent but contextually probable (*got cleats at the sports store*), corrective focus differs significantly from wide focus; new-information focus does not (Figure 3, 'Low Freq + High Prob'). Finally, neither type of narrow focus differs from wide focus when it is an infrequent word focused in an improbable context (*got toys at the sports store*, Figure 4, 'Low Freq + Low Prob').

The analysis of **f0 ranges** finds parallel patterns to the above results of f0 shapes. There are no significant differences in f0 ranges when the Pre-Focus and Focus intervals are analyzed either jointly (i.e. treated as one region) or separately (t's < 1.723, p's > 0.086). The interaction between word frequency, contextual probability and focus types appears when the Post-Focus interval is analyzed alone or jointly with the Focus interval. In the condition of *lexically frequent and contextually probable words,* both types of narrow focus have significantly larger f0 ranges than wide focus (t's > 2.524, p's < 0.05, except for new-information focus in the Post-Focus interval: $t = 1.458$; $p = 0.147$). In the condition of *lexically frequent but contextually improbable words*, only new-information focus has larger f0 ranges than wide focus (t's > 1.994, p's < 0.05); corrective focus does not (t's < 1.650, p's > 0.100). In contrast, for *low-frequency but high-probability words*, corrective focus has larger f0 ranges than wide focus (t's > 2.159, p's < 0.05); new-information focus patterns with wide focus (t's < 1.091, p's > 0.276). Lastly, neither type of narrow focus differs from wide focus when *low-frequency words are focused in low-probability contexts* (t's < 1.366, p's > 0.173).[4]

---

4 Here we do not report statistics for the Focus and Post-Focus intervals separately, due to reasons of readability, and more importantly because we do not think that this distinction (i.e. whether it is the Focus or Post-Focus interval

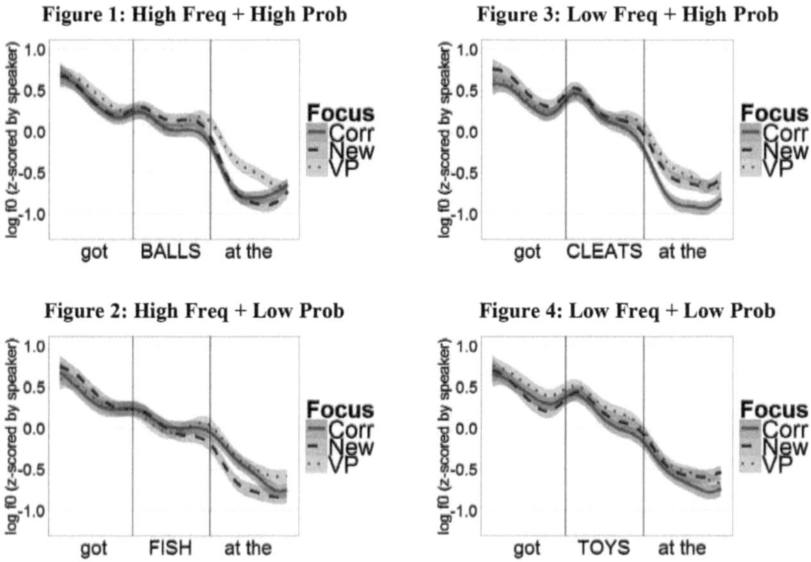

Figures 1–4: Best-fitted curves with 95% confidence intervals for the f0 values (in semitone, standardized by speaker) in the pre-focus, focus and post-focus regions of an utterance.

As a whole, we find that narrow focus brings greater prosodic prominence than wide focus, but this effect disappears under certain conditions of word frequency and contextual probability. Specifically, narrow corrective focus differs from wide focus only when the word carrying corrective information in narrow focus is probable in its sentence context. Conversely, new-information focus differs from wide focus when the word carrying new information in narrow focus is a frequent word. This suggests that the prosodic prominence associated with information structure is modulated by word frequency and contextual probability.

which shows significant differences) is theoretically relevant for the claims we are making in this paper.

4. Individual results

In the previous section, we summarized the overall patterns when all participants are investigated as a group. Let us now explore whether and how individual participants differ from one another. In this section, we will first look at the overall prosody of individual speakers, focusing on f0 shapes and ranges (section 4.1.). Then, we will examine the different experimental conditions, to see how individual speakers produce different types of focus in different conditions of word frequency and contextual probability (section 4.2.).

4.1. Overall prosody of utterances

Overall, in terms of general prosodic patterns, speaker-specific variation occurs both qualitatively and quantitatively. Between-subject variability and within-subject consistency were observed in both the shapes of f0 contours and the ranges of f0 values.

First, *the shapes of f0 contours* vary greatly from participant to participant. In a given condition, participants differ in the number, locations and relative height of the f0 peaks and valleys that they produce in an utterance. To illustrate the extent of variability, we plotted a sample of five participants whose f0 shapes are clearly distinct from one another. Figure 5 shows the observed f0 contours produced by these participants for new-information, frequent words that are narrowly-focused in probable contexts (e.g., *What did Rachel and Carolyn get at the sports store? They got balls at the sports store.*) We can see that participants 04 (triangles), 06 (dots) and 09 (dashes) all tend to produce a high tone on the focused word (i.e. *balls*) – thus showing overall consistency in this regard. However, their choices regarding the adjacent tones differ. Participant 06's utterances on average have a low tone preceding the high tone, participant 04's in general have another high tone preceding the high tone, and participant 09's seem to have a low tone following the high tone. Furthermore, participant 01 (squares) and participant 04 both show a clear tendency of declination, but participant 04's utterances have two high tones whereas participant 01's do not have apparent tone targets. Lastly, participant 07 (solid line) distinctively produces the focused word with a low tone. Such diversity is found among other participants and in other conditions as well.

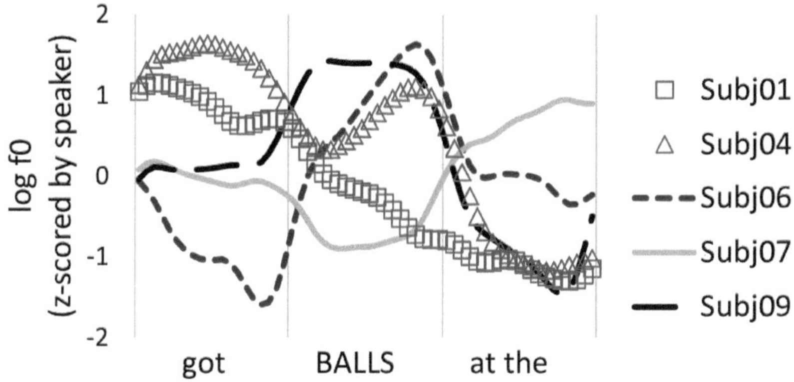

Figure 5: Observed mean f0 (in semitone, standardized by speaker) for participants 01, 04, 06, 07 and 09 in the narrow new-information focus, high word frequency and high contextual probability condition.

Although different participants produce different shapes of f0 contours, they show consistent patterns within their own utterances. For example, Figure 6 provides a glance at the observed f0 contours produced by participant 04 in all twelve experiment conditions. We can see that participant 04's utterances are quite similar to one another, regardless of the condition. To further illustrate the intra-subject consistency with better graphical legibility, Figure 7 shows the smoothing spline ANOVA results of three individual participants, including participant 04, in all four information-theoretic conditions. These three participants were chosen because they had strong preferences regarding f0 shapes. We can see that participant 01's utterances (top row) mostly follow a declination slope, although a low tone occasionally occurs around the end of the Focus interval. Participant 04 (middle row) consistently produces a high tone in the Pre-Focus interval and another high tone, downstepped, in the Focus interval, except there is sometimes a low tone proceeding and/or following the second high tone. Participant 06 (bottom row) generally produces a low tone in the Pre-Focus interval and a high tone in the Focus interval, which is often followed by another low tone. Speaker-specific preferences of this sort are also found for most of the other participants in our data.

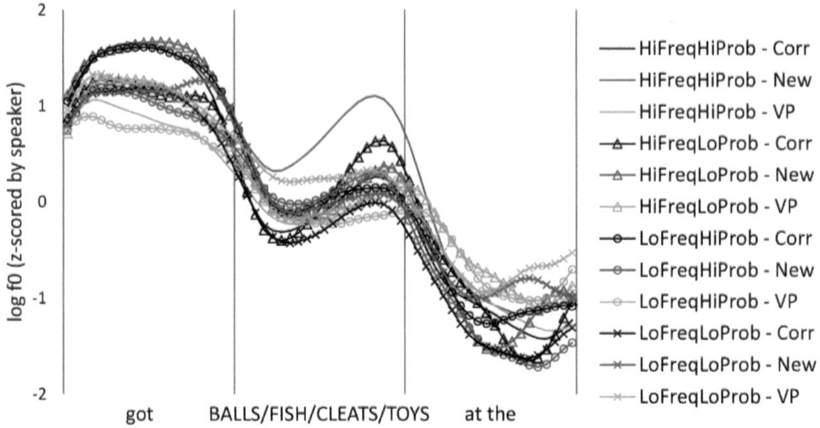

Figure 6: Observed mean f0 (in semitone, standardized by speaker) of participant 04 in all the experiment conditions.

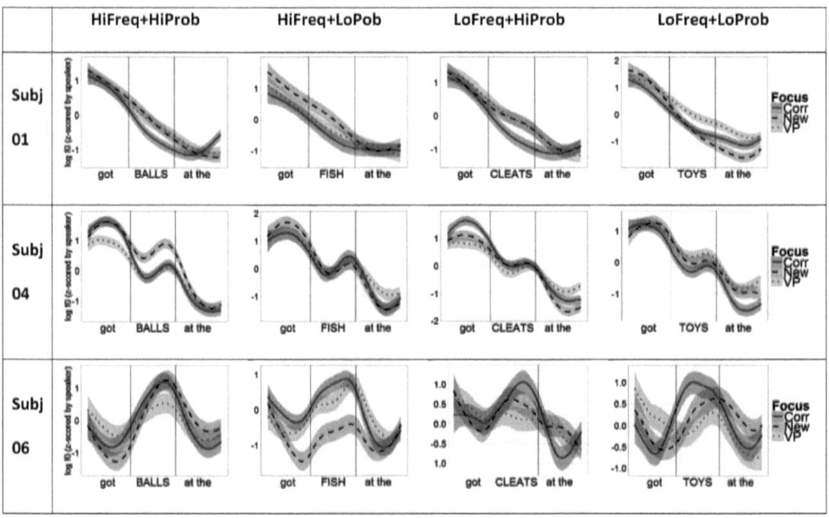

Figure 7: Best-fitted curves with 95% confidence intervals for the f0 values (in semitones, standardized by speaker) produced by participants 01, 04 and 06.

We also find speaker-specific effects in *the ranges of f0 values*. Some participants regularly employ a large proportion of their f0 register, while others regularly employ a small proportion of their f0 register. To illustrate, let us take a close look at the sentence region from the Pre-Focus interval to the Post-Focus interval. Figure 8 shows the average f0 ranges with 95% confidence intervals (1.96 standard errors) produced by individual participants. We can see that every participant differs from some other participant(s). Pairwise comparisons with the Bonferroni adjustment show that, between the sixteen participants, everyone significantly differs from at least two other people and as many as thirteen other people (p's < 0.05). For example, participant 05, whose f0 ranges are largest on average (mean = 2.787) and the least variable among all participants (standard deviation = 0.512), differs from participants 01, 03, 04, 06, 07, and 09–16. On the other hand, participant 12, whose f0 ranges are smallest on average (mean = 1.582), differs from participants 02, 04, 05, 06, 08, 09, 11 and 16. Even participant 07, whose f0 ranges are the most variable among all participants (standard deviation = 1.253), differs from participants 02, 05, 08 and 11 (by being smaller). More details about other participants can be observed in Figure 8.

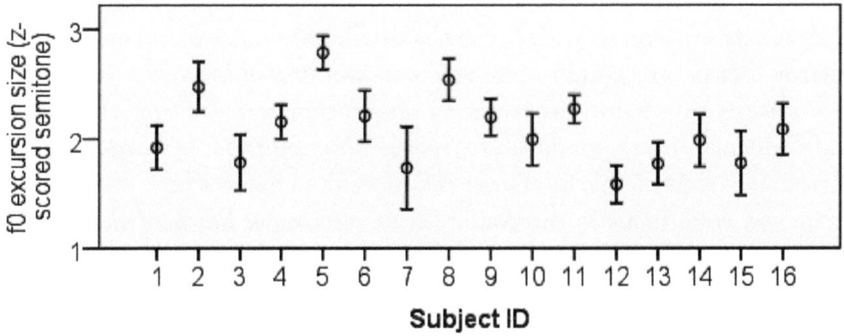

Figure 8: The observed f0 ranges (calculated from semitones standardized by speaker) with 95% confidence intervals for individual participants in the sentence region from the pre-focus interval to the post-focus interval. A larger f0 range indicates that the speaker employs a bigger proportion of his/her f0 register for this sentence region.

In sum, individual participants appear to be fairly different from one another, yet consistent within one's own utterances, in terms of the f0 shapes they adopt and how large a proportion of their f0 register they use. This suggests evidence for speaker-specific behavior in the overall prosodic patterns of utterances and the extent to which people utilize their vocal capacity to produce prosodic cues.

4.2. Prosodic encoding of informativity

Now that we have seen speaker-specific effects on the overall shapes and ranges of f0, let us move on to the individual differences in how their prosody reflects the informativity of linguistic elements. Since a given participant's f0 shapes are similar across the conditions, i. e., different types of focus and different levels of word frequency and contextual probability (see section 4.1.), only f0 ranges are of the interest in this subsection. To draw a direct comparison between the group trends and the individual patterns, we present the results of the sentence region from the Focus interval to the Post-Focus interval, where the group analysis finds significant differences (see section 3.).

First, we observe some between-subject variability in terms of the *direction* of distinctions between different kinds of information. As presented in section 3., there are three main patterns when all sixteen participants are analyzed as a group: (i) wide focus has smaller f0 ranges than both types of narrow focus in the condition of *frequent and probable words*, (ii) narrow new-information focus has larger f0 ranges than narrow corrective focus and wide focus in the condition of *frequent but improbable words*, and (iii) narrow corrective focus has larger f0 ranges than narrow new-information focus and wide focus in the condition of *infrequent but probable words*. The analysis of individual participants finds each pattern in eight or nine people out of sixteen: pattern (i) is exhibited by participants 01, 02, 04, 05, 07, 10, 12, 14 and 15; pattern (ii) is exhibited by participants 01, 02, 06, 07, 12, and 14–16; pattern (iii) is exhibited by participants 04 and 07–13. In other words, only about half of the participants conform to the group trends regarding how information-structural types are differentiated in a given information-theoretic condition, and it is not the same individuals in every condition. However, it is worth noting that there are no alternative 'competitor' patterns – instead, the participants who do not match the overall group trends show a mix of patterns in the different conditions. Thus, although the overall group trends (as summarized in (i)-(iii) above) are not exhibited by everyone, they nevertheless constitute the clearest patterns that emerge from the data.

Participants also differ in the *magnitude* of the information-structural distinctions they make. To illustrate, Figure 9 shows the f0 ranges of individual participants in the condition of high word frequency and high contextual probability. It appears that some people make clearer distinctions than others. For example, the differences between wide and narrow focus are bigger in participants 07 and 15 than participants 04 and 14. Participants 07 and 15 use substantially larger f0 ranges for the utterances containing narrow focus than the utterances containing wide focus, whereas participants 04 and 14 differentiate these two kinds of utterances to a lesser degree. Similarly variable patterns are found in other conditions as well.

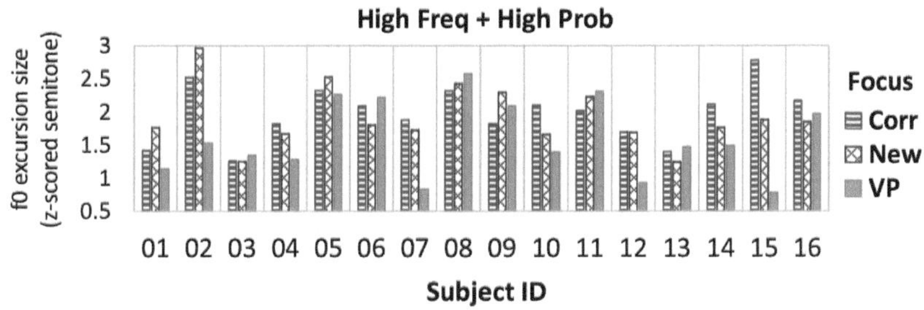

Figure 9: The observed f0 ranges (calculated from semitones standardized by speaker) in the sentence region from the Focus interval to the Post-Focus interval for individual participants in the condition of high word frequency and high contextual probability. A larger f0 range indicates that a bigger proportion of the speaker's f0 register is employed.

Let us now consider how internally-consistent speakers are in terms of the (i) directionality and (ii) magnitude of the information-structural distinctions that they produce. We find considerable trial-by-trial variation in the direction of the information-structural distinctions produced by individual participants (although the patterns reach significance in the group analysis). Particularly, there is little indication of interactions between speaker (i.e. who is speaking) and any of the informativity factors in terms of the *direction* of distinctions between different kinds of information. In other words, the overall group results also hold on the level of individual speakers, and it is generally not the case that, depending who the speaker is, one particular type of information would consistently lead to smaller (or larger) f0 ranges than another particular type of information.

Interestingly, if we look at the magnitude of these distinctions, we find more speaker-internal consistency. Some participants regularly produce much larger f0 ranges for one type of focus than another, while some others regularly produce only slightly larger f0 ranges for one type of focus than another. For example, let us take a close look at the participants who conform to more than one group trend: participants 01, 02, 04, 07, 10, 12, 14 and 15. It appears that they can be divided into two subgroups such that, across information-theoretic conditions, one subgroup consistently produces stronger cues for information-structural distinctions than the other subgroup. To illustrate, Figure 10 shows the differences in f0 ranges produced by the eight participants in the information-theoretic conditions where they conform to the group trends regarding the information-structural distinctions. These differences were calculated with respect to the group trend in each condition, i. e. patterns (i-iii) that we summarized towards the beginning of this subsection. Specifically, the bars for the high-frequency high-probability condition represent the differences between wide focus and the other two types focus (i. e. the f0 range in wide focus subtracted from the f0 ranges in new-information narrow focus and corrective narrow focus) based on pattern (i)), the bars for the high-frequency low-probability condition represent the differences between narrow new-information focus and the other two types of focus (i. e. the f0 ranges in wide focus and corrective focus subtracted from the f0 range in new-information narrow focus, based on pattern (ii)), and the bars for the low-frequency high-probability condition represent the differences between narrow corrective focus and the other two types of focus (i. e. the f0 ranges in wide focus and new-information focus subtracted from the f0 range in corrective narrow focus, based on pattern (iii)). For the participants who do not conform to all of these three group patterns (i. e. participants 01, 02, 04, 10, 14, and 15), we only calculated the differences in f0 ranges for the information-theoretic conditions where they do. Thus, we can see that participants 02, 07, 12 and 15 produce pattern (i) with larger differences than participants 01, 04, 10 and 14, participants 02, 07, 12 and 15 produce pattern (ii) with larger differences than participants 01 and 14, and participants 07 and 12 produce pattern (iii) with larger differences than participants 04 and 10. Essentially, in a given information-theoretic condition, participants 02, 07, 12 and 15 consistently use larger differences in f0 ranges than participants 01, 04, 10 and 14 for a given direction of information-structural distinctions. In general, this observation leads us to speculate that what matters (in terms of encoding and perceiving informativity) are not the absolute but rather the relative values.

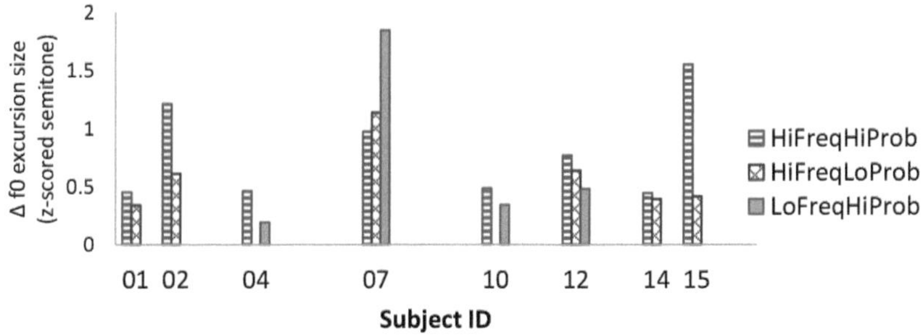

Figure 10: The observed differences in f0 ranges (calculated from semitones standardized by speaker) in the sentence region from the Focus interval to the Post-Focus interval for individual participants who conform to more than one group trend. The differences were calculated based on the group trend in each condition.

To sum up, when we look at individual differences in how speakers encode informativity prosodically, we find that about half of the speakers clearly exhibit the f0 range patterns that we observed for the group as a whole in terms of which conditions have larger vs. smaller f0 ranges, and the remaining speakers show more variable data. In terms of the magnitude of their f0 ranges, speakers are largely internally consistent, and our data suggests that speakers differ in how much they modulate f0 to signal informativity. Broadly speaking, this suggests that what matters in terms of encoding information-theoretic notions prosodically are relative, not absolute, values – an observation which is in line with prior work on prosody and information structure.

5. General discussion

Our experiment investigates how information structure and information-theoretic properties interact in shaping the prosody of an utterance and how individual speakers differ in the overall prosody of utterances and the prosodic encoding of informativity. Existing studies have examined prosody from information-theoretic and information-structural perspectives, but the interaction between these two kinds of informativity factors has not been thoroughly investigated. In addition, prior work mostly focuses on the general trends among speakers, and little has been said about the differences between or within speakers. A better understanding of these issues is important because they are involved in fundamental questions regarding the functions and constraints of the prosodic system. In this section, we discuss how our results relate to these issues, and what their broader implications are.

Our results in section 3. show that, when the **participants are analyzed as a whole,** the prosodic effects of information structure are modulated by information-theoretic factors. In particular, we find differential effects of contextual probability and word frequency on corrective narrow focus vs. new-information narrow focus. *Corrective narrow focus* results in significant f0 movement only when the word carrying corrective information is *probable* in the context. However, *new-information narrow focus* results in significant f0 movement only when the word carrying new information is a *frequent* word. When the narrowly focused word is lexically frequent and contextually probable, both types of narrow focus have greater f0 movement than wide focus. In contrast, when the narrowly focused word is infrequent and improbable, neither type of narrow focus type is distinguishable from wide focus. This fits with our prediction that the prosodic prominence associated with information structure would be weakened when other factors also demand prosodic prominence.

Taken together, these findings pose a challenge to the widespread view that narrow focus is (consistently) associated with greater prosodic prominence than wide focus. In fact, prior work on the phonetic realization of information structure suggests a prominence hierarchy, such that contrastive/corrective information is prosodically more marked than 'plain' new information, and new information in narrow focus is prosodically more marked than new information in wide focus (e.g., English: Breen et al., 2010; Katz and Selkirk 2011; German: Baumann et al., 2006; Mandarin Chinese: Ouyang and Kaiser, 2015; Xu, 1999). To the contrary, we did not see this hierarchy in our data – we found that contextual probability and word frequency need to be considered in order to understand the relative prosodic prominence of different focus types. Interestingly, it seems that many existing studies have focused on relatively probable contexts and have not manipulated word frequency, which may explain the hierarchical relation previously found between corrective focus, new-information focus and wide focus (i.e. narrow corrective > narrow new > wide new).

Consider a hypothetical study that has a mix of high-frequency and low-frequency words focused in probable contexts. Based on our results, in such a study: (a) corrective focus will have greater prominence than wide focus, since the contexts are probable, and (b) new-information focus will be less prominent than corrective focus and more prominent than wide focus, because frequent words pattern with the former but infrequent words pattern with the latter. These predictions are confirmed by a follow-up analysis where we pooled the conditions of word frequency and excluded the condition of high contextual probability. Using the approaches described in section 2.3., we found significant differences in the Focus and Post-Focus intervals. The f0 movement was largest for corrective focus, second largest for new-information focus, and smallest for wide/VP focus. In other words, the common generalization about the prominence hierarchy between the three types of focus might be an epiphenomenon stemming from not controlling word frequency and using relatively probable contexts.

Here we will not further discuss why a word's information-theoretic properties interact with its information-structural status in the particular way we observed, since it is not the focus of this paper. Nevertheless, our findings highlight the importance of disentangling information structure and information-theoretic factors. To fully understand how prosody encodes informativity, it is necessary to integrate the work in the information-theoretic approach and the work in the information-structural approach (see Wagner and Watson, 2010: 933, for relevant discussion).

Let us now consider the **nature and extent of individual variation.** In this section, we will consider the shapes of f0 contours, the ranges of f0 values, the directionality of differences in f0 ranges (i.e. which conditions have larger/smaller f0 ranges than other conditions), and the magnitude of differences in f0 ranges (i.e. how much larger/smaller the f0 ranges are in one condition than another). As we saw in section 4., if we look at the overall prosody and f0 ranges that speakers produce, abstracting away from informativity notions, we find that speakers differ from one another but are internally quite consistent. In other words, individual speakers have preferences with regard to the shapes of f0 contours and the ranges of f0 values, generally speaking. Then, when we look at how individual speakers encode informativity notions prosodically, we find that the group-level patterns regarding the directionality of information-structural distinctions are exhibited by many, but not all, speakers. Interestingly, when we look more closely at how internally consistent speakers are in this regard, we find that speakers show considerable internal variation in the directionality of distinctions they produce (i. e. whether a particular type of focus has larger or smaller f0 ranges than another particular type of focus). In contrast, in terms of the magnitude of distinctions they produce (i. e. how much larger or smaller the f0 ranges are in one particular type of focus than another), speakers are more internally consistent while, again, different from one another. Nevertheless, the group patterns are statistically significant (in analyses that include subjects and items as random factors), and thus we conclude that they are still meaningful even in the face of individual variation.

As we noted in section 3., the group analysis reveals three main patterns which highlight the interplay of information theory and information structure: (i) wide focus has smaller f0 ranges than both types of narrow focus in the condition of *frequent and probable words*, (ii) narrow new-information focus has larger f0 ranges than narrow corrective focus and wide focus in the condition of *frequent but improbable words*, and (iii) narrow corrective focus has larger f0 ranges than narrow new-information focus and wide focus in the condition of *infrequent but probable words*. We found that about half of the participants clearly exhibit these patterns. Importantly, there is no other 'competitor pattern' that emerges from the data, as the rest of the participants exhibit more than one other pattern (e. g. some make corrective focus the least prominent while others make new-information focus the least prominent, as can be seen in Figure 9).

Thus, we observe a set of patterns that a large subset of participants exhibits, and then other, seemingly highly variable, non-systematic patterns. It seems that speakers loosely follow principles determined by information-theoretic factors and information structure, and collectively show a systematic relationship between prosodic prominence and informativity. A related phenomenon has been found in the field of speech processing. Studies on accent prediction have argued that using speaker-dependent parameters does not substantially improve a model's performance in predicting whether a word receives an accent or not, because the variability in placing an accent or not between speakers is similar to that within a speaker (Badino and Clark, 2007; Shriberg et al., 1996; Yuan et al., 2005). Our results are consistent with these findings.

While the directions of differences in f0 ranges are closely tied to informativity factors, some other aspects of f0 – including the ranges of f0 values, the sizes of differences in f0 ranges, and the shapes of f0 contours – appear to show speaker-specific behavior. Given the multi-functionality of prosody, it is not surprising that these other f0 parameters do not supply strong cues for the particular factors we investigated. In terms of the range of f0 values in an utterance and the magnitude of fluctuations in f0 ranges across utterances, prior work has found that these aspects of f0 ranges can reflect the speaker's emotions and psychological traits. For example, sad, depressed, anxious, irritated, tense or fearful speech employs more limited f0 ranges than happy or angry speech (e.g. Johnstone and Scherer, 1999; Morley et al., 2011). Furthermore, children and young adults with autistic spectrum disorders use more exaggerated f0 ranges than individuals with typical development (e.g. Hubbard and Trauner, 2007; Paul et al., 2008; Sharda et al., 2010). Thus, it is likely that the speaker-specific patterns regarding f0 ranges observed in this study correlate with individual participants' mood or personal characteristics.

Similarly, f0 shapes have been shown to convey many other kinds of pragmatic meanings that are not investigated in this study, such as the speaker's beliefs or the relationship between an utterance and a subsequent one (e.g. Pierrehumbert and Hirschberg, 1990; Ward and Hirschberg, 1985, 1986). Due to the nature of our experiment (i.e. reading aloud sentence pairs), the stimuli were underspecified in these aspects and open for the participant's own interpretations. Therefore, the presence of speaker-specific patterns in f0 shapes might imply that individual speakers have preferences regarding how to fill in unspecified details at the pragmatic level. This is an interesting question that would benefit from future work.

Thus, based on our results, it appears that f0 shapes are less informative than f0 ranges in distinguishing the three information-structural types of interest, namely corrective narrow focus, new-information narrow focus, and wide focus. F0 shapes differentiate these three types of focus when we look at all speakers as a whole, but not when we look at each speaker individually. In contrast, the directions of differences in f0 ranges distinguish focus types at *both* the group level and the individual level. This suggests that f0 ranges may have a greater contribution than f0 shapes to the prosodic marking of information structure. We leave this question open for future work.

In sum, this study contributes to our understanding of individual differences, providing empirical evidence for inter- and intra-speaker variability in the prosodic encoding of informativity. Our results are consistent with previous observations that prosody can exhibit speaker-specific behavior. Furthermore, we show that apparent differences among the participants in a study do not necessarily constitute stable speaker-specific patterns. Instead, the prosodic dimensions that do not show participants' individual preferences may be the key dimensions that reflect the linguistic distinctions in question (e.g. the direction of differences in f0 ranges in this study). In addition, we discuss possible explanations for speaker-specific behavior in the prosodic dimensions we investigate. Prosody appears to be highly multi-functional and tolerant of idiosyncrasies to a considerable extent.

6. Conclusions

On the basis of the psycholinguistic production study reported in this paper, we can draw three main conclusions. *First*, information structure and information-theoretic factors interact in influencing an utterance's prosody. Our results show that word frequency modulates the prosodic effect of new-information focus (see also Baker and Bradlow, 2009), whereas contextual probability modulates the prosodic effect of corrective focus. *Second*, our findings suggest the presence of speaker-specific behavior in prosody. Speakers have individual preferences regarding the prosodic patterns of utterances and the magnitude of prosodic cues for informativity. *Third*, we did not see signs of speaker-specific behavior in the directions of prosodic distinctions between information categories – in other words, this seems to be a key dimension where English speakers show consistent behavior in terms of how informativity related factors are encoded in prosody. In sum, this work contributes to our understanding of prosody by providing empirical evidence for the interaction between word frequency and new-information focus, the interaction between contextual probability and corrective focus, as well as the nature and extent of speaker-specific variation. Our findings highlight the importance of disentangling information structure and information-theoretic factors and examining both inter- and intra-speaker variability.

Acknowledgements

Earlier version of this work was presented at the 4th *International Summer School 2013 on Speech Production and Perception: Speaker-Specific Behavior*, the 27th Annual CUNY Conference on Human Sentence Processing (2014, Columbus, Ohio, USA), the 38th Annual Penn Linguistics Conference in 2014, and the 36th Annual Meeting of the Cognitive Science Society (2014, Quebec City, Quebec, Canada). We thank the audience members for their valuable comments and suggestions. Thanks also go to the USC Language Processing Lab group for feedback during the development of this project. Last but not least, we thank the editors of this book and three anonymous reviewers, whose comments and suggestions greatly enhanced this chapter.

References

Allen, J.S., Miller, J.L., and DeSteno, D. (2003). Individual talker differences in voice-onset-time. *The Journal of the Acoustical Society of America, 113*, 544–552.

Andreeva, B., Barry, W.J., and Steiner, I. (2007). Producing phrasal prominence in German. In *Proceedings of the 16th International Congress of Phonetic Sciences*, 1209–1212.

Aylett, M., and Turk, A. (2004). The smooth signal redundancy hypothesis: A functional explanation for relationships between redundancy, prosodic prominence and duration in spontaneous speech. *Language and Speech, 47*, 31–56.

Badino, L., and Clark, R.A.J. (2007). Issues of optionality in pitch accent placement. In *Proceedings of the 6th ISCA Workshop on Speech Synthesis*, 252–257.

Baker, A. (2006). Quantifying diphthongs: A statistical technique for distinguishing formant contours. *Paper presented at New Ways of Analyzing Variation (NWAV) 35*, Columbus, OH.

Baker, R. E., and Bradlow, A.R. (2009). Variability in word duration as a function of probability, speech style, and prosody. *Language and Speech, 52*, 391–413.

Bates, D., Maechler, M., Bolker, B., and Walker, S. (2014). *lme4: Linear mixed-effects models using Eigen and S4*. R package version 1.1–7.

Baumann, S., Grice, M., and Steindamm, S. (2006). Prosodic marking of focus domains-categorical or gradient. In *Proceedings of Speech Prosody 2006*, 301–304.

Bell, A., Brenier, J.M., Gregory, M., Girand, C., and Jurafsky, D. (2009). Predictability effects on durations of content and function words in conversational English. *Journal of Memory and Language, 60*, 92–111.

Bell, A., Jurafsky, D., Fosler-Lussier, E., Girand, C., Gregory, M., and Gildea, D. (2003). Effects of disfluencies, predictability, and utterance position on word form variation in English conversation. *The Journal of the Acoustical Society of America, 113*, 1001–1024.

Breen, M., Fedorenko, E., Wagner, M., and Gibson, E. (2010). Acoustic correlates of information structure. *Language and Cognitive Processes, 25*, 1044–1098.

Brown, G. (1983). Prosodic structure and the given/new distinction. In A. Cutler and D. Robert Ladd (eds.), *Prosody: Models and measurements*. Springer Science and Business Media.

Brysbaert, M., and New, B. (2009). Moving beyond Kucera and Francis: A critical evaluation of current word frequency norms and the introduction of a new and improved word frequency measure for American English. *Behavior Research Methods, 41*, 977–990.

Calhoun, S. (2010). How does informativeness affect prosodic prominence? *Language and Cognitive Processes, 25*, 1099–1140.

Chen, Y., and Braun, B. (2006). Prosodic realization of information structure categories in Standard Chinese. In *Proceedings of Speech Prosody 2006*.

Ching, M. K. L. (1982). The question intonation in assertions. *American Speech*, 95–107.

Cho, T., and Keating, P.A. (2001). Articulatory and acoustic studies on domain-initial strengthening in Korean. *Journal of Phonetics, 29*, 155–190.

Clopper, C.G., and Pierrehumbert, J.B. (2008). Effects of semantic predictability and regional dialect on vowel space reduction. *The Journal of the Acoustical Society of America, 124*, 1682–1688.

Cooper, W.E., Eady, S.J., and Mueller, P.R. (1985). Acoustical aspects of contrastive stress in question-answer contexts. *The Journal of Acoustical Society of America, 77*, 2142–2156.

Couper-Kuhlen, E. (1984). A new look at contrastive intonation. In R. J. Watts and U. Weidman (eds.), *Modes of interpretation: Essays presented to Ernst Leisi on the occasion of his 65th Birthday*. Tübingen: Gunter Narr Verlag.

Dahan, D., and Bernard, J.M. (1996). Interspeaker variability in emphatic accent production in French. *Language and Speech, 39*, 341–374.

Davidson, L. (2006). Comparing tongue shapes from ultrasound imaging using smoothing spline analysis of variance. *The Journal of the Acoustical Society of America, 120*, 407–415.

Dik, S.C. (1997). *The theory of functional grammar*. Berlin: Mouton De Gruyter.

Eady, S.J., and Cooper, W.E. (1986). Speech intonation and focus location in matched statements and questions. *The Journal of the Acoustical Society of America, 80*, 402–415.

Ferguson, S. H. (2004). Talker differences in clear and conversational speech: Vowel intelligibility for normal hearing listeners. *The Journal of the Acoustical Society of America 116*, 2365–2373.

Ferguson, S. H., and Kewley-Port, D. (2007). Talker differences in clear and conversational speech: Acoustic characteristics of vowels. *Journal of Speech, Language, and Hearing Research, 50*, 1241–1255.

Fougeron, C., and Keating, P.A. (1997). Articulatory strengthening at edges of prosodic domains. *The Journal of the Acoustical Society of America, 101*, 3728–3740.

Fowler, C. A., and Housum, J. (1987). Talkers' signaling of "new" and "old" words in speech and listeners' perception and use of the distinction. *Journal of Memory and Language, 26*, 489–504.

Garcia, D. (2010). Robust smoothing of gridded data in one and higher dimensions with missing values. *Computational Statistics and Data Analysis, 54*, 1167–1178.

Gregory, M. L., Raymond, W.D., Bell, A., Fosler-Lussier, E., and Jurafsky, D. (1999). The effects of collocational strength and contextual predictability in lexical production. *Chicago Linguistic Society, 35*, 151–166.

Gu, C. (2002). *Smoothing spline ANOVA models*. New York: Springer.

Gu, C. (2014). *gss: General smoothing splines*. R package version 2.1-4.

Gussenhoven, C. (1983). Testing the reality of focus domains. *Language and Speech, 26*, 61–80.

Jennifer, J., Warren, P., and Drager, K. (2006). Factors influencing speech perception in the context of a merger-in-progress. *Journal of Phonetics, 34*, 458–484.

Hubbard, K., and Trauner, D.A. (2007). Intonation and emotion in autistic spectrum disorders. *Journal of Psycholinguistic Research, 36*, 159–173.

Johnstone, T., and Scherer, K.R. (1999). The effects of emotions on voice quality. In *Proceedings of the 16th International Congress of Phonetic Sciences*, 2029–2032.

Katz, J., and Selkirk, E. (2011). Contrastive focus vs. discourse-new: Evidence from phonetic prominence in English. *Language, 87*, 771–816.

Kawahara, H., de Cheveigne, A., and Patterson, R.D. (1998). An instantaneous-frequency-based pitch extraction method for high-quality speech transformation: Revised TEMPO in the STRAIGHT-suite. In *Proceedings of the 5th International Conference on Spoken Language Processing*, 1367–1370.

Krahmer, E., and Swerts, M. (2001). On the alleged existence of contrastive accents. *Speech Communication, 34*, 391–405.

Krivokapić, J., and Byrd, D. (2012). Prosodic boundary strength: An articulatory and perceptual study. *Journal of Phonetics, 40*, 430–442.

Kuznetsova, A., Brockhoff, P.B., and Bojesen Christensen, R.H. (2015). *lmerTest: Tests in Linear Mixed Effects Models*. R package version 2.0-25.

Ladd D. R. (1996). *Intonational phonology*. Cambridge: Cambridge University Press.

Lieberman, P. (1963). Some effects of semantic and grammatical context on the production and perception of speech. *Language and Speech, 6*, 172–187.

Loakes, D., and McDougall, K. (2010). Individual variation in the frication of voiceless plosives in Australian English: A study of twins' speech. *Australian Journal of Linguistics, 30*, 155–181.

Morley, E., van Santen, J., Klabbers, E., and Kain, A. (2011). F0 range and peak alignment across speakers and emotions. In *Proceedings of 2011 IEEE International Conference on Acoustics, Speech and Signal Processing*, 4952–4955.

Munson, B., and Pearl Solomon, N. (2004). The influence of phonological neighborhood density on vowel articulation. *Journal of Speech, Language, and Hearing Research, 47*, 1048–1058.

Niebuhr, O., D'Imperio, M., Gili Fivela, B., and Cangemi, F. (2011). Are there "shapers" and "aligners"? Individual differences in signalling pitch accent category. In *Proceedings of the 17th International Congress of Phonetic Sciences*, 120–123.

Nolan, F. (2003). Intonational equivalence: an experimental evaluation of pitch scales. In *Proceedings of the 15th International Congress of Phonetic Sciences*, 771–774.

Ouyang, I.C., and Kaiser, E. (2015). Prosody and information structure in a tone language: An investigation of Mandarin Chinese. *Language, Cognition and Neuroscience,* 30, 57–72.

Ouyang, I.C., and Kaiser, E. (2014). Prosodic encoding of informativity: Word frequency and contextual probability interact with information structure. In *Proceedings of the 36th Annual Meeting of the Cognitive Science Society*, 1120–1125.

Pan, S., and Hirschberg, J. (2000). Modeling local context for pitch accent prediction. In *Proceedings of the 38th Annual Conference of the Association for Computational Linguistics*, 233–240.

Paul, R., Bianchi, N., Augustyn, A., Klin, A., and Volkmar, F.R. (2008). Production of syllable stress in speakers with autism spectrum disorders. *Research in Autism Spectrum Disorders,* 2, 110–124.

Pierrehumbert, J.B., and Hirschberg, J. (1990). The meaning of intonational contours in the interpretation of discourse (pp. 271–311), In P.R. Cohen, J.L. Morgan, and M. E. Pollack (eds.), Intentions in communication. Cambridge, Massachusetts: MIT Press.

Pitrelli, J. F. (2004). ToBI prosodic analysis of a professional speaker of American English. In *Proceedings of Speech Prosody 2004.*

Pluymaekers, M., Ernestus, M., and Baayen, R.H. (2005a). Lexical frequency and acoustic reduction in spoken Dutch. *The Journal of the Acoustical Society of America, 118*, 2561–2569.

Pluymaekers, M., Ernestus, M., and Baayen, R.H. (2005b). Articulatory planning is continuous and sensitive to informational redundancy. *Phonetica, 62*(2–4), 146–159.

Prince, E. (1992). The ZPG letter: Subjects, definiteness, and information-status. (pp. 295—325), In William C. Mann and Sandra A. Thompson (Eds.), *Discourse description: Diverse analyses of a fund-raising text.* Philadelphia: John Benjamins.

Rietveld, T., Kerkhoff, J., and Gussenhoven, C. (2004). Word prosodic structure and vowel duration in Dutch. *Journal of Phonetics, 32*, 349–371.

Rooth, M. (1992). A theory of focus interpretation. *Natural Language Semantics, 1*, 75–116.

Scarborough, R. (2010). Lexical and contextual predictability: Confluent effects on the production of vowels. *Laboratory Phonology, 10*, 557–586.

Selkirk, E. (1984). *Phonology and syntax: The relation between sound and structure.* Cambridge, Massachusetts: MIT Press.

Sharda, M., Subhadra, T.P., Sahay, S., Nagaraja, C., Singh, L., Mishra, R., Sen, A., Singhal, N., Erickson, D., and Singh, N.C. (2010). Sounds of melody—Pitch patterns of speech in autism. *Neuroscience Letters, 478*, 42–45.

Shriberg, E., Ladd, D.R., Terken, J. and Stolcke, A. (1996). Modeling pitch range variation within and across speakers: Predicting f0 targets when "speaking up". In *Proceedings of the 4th International Conference on Spoken Language Processing*, 1–4.

Shue, Y.-L., Keating, P., Vicenik, C., and Yu, K. (2011). VoiceSauce: A program for voice analysis. In *Proceedings of the 17th International Congress of Phonetic Sciences*, 1846–1849.

Smith, R., and Hawkins, S. (2012). Production and perception of speaker-specific phonetic detail at word boundaries. *Journal of Phonetics, 40*, 213–233.

Theodore, R.M., Miller, J.L., and DeSteno, D. (2007). The effect of speaking rate on voice-onset-time is talker-specific. In *Proceedings of the 16th International Congress of Phonetic Sciences*, 473–476.

Trouvain, J., and Grice, M. (1999). The effect of tempo on prosodic structure. In *Proceedings of the 14th International Congress of Phonetic Sciences*, 1067–1070.

Vallduví, E., and Vilkuna, M. (1998). On rheme and kontrast. In P. W. Culicover and L. McNally (eds.), *Syntax and semantics 29: The Limits of Syntax*. San Diego: Academic Press.

Van Donzel, M.E., and Koopmans-van Beinum, F.J. (1996). Pausing strategies in discourse in Dutch. In *Proceeding of Fourth International Conference on Spoken Language Processing (ICSLP '96)*, 1029–1032.

Van Son, R. J. J. H., Koopmans-van Beinum, F.J., and Pols, L.C.W. (1998). Efficiency as an organizing principle of natural speech. In *Proceedings of the 5th International Conference on Spoken Language Processing (ICSLP '98)*, 2375–2378.

Wagner, M., and Watson, D.G. (2010). Experimental and theoretical advances in prosody: A review. *Language and Cognitive Processes, 25*, 905–945.

Ward, G. L., and Hirschberg, J. (1985). Implicating uncertainty: The pragmatics of fall-rise intonation. *Language*, 747–776.

Ward, G. L., and Hirschberg, J. (1986). Reconciling uncertainty with incredulity: A unified account of the L*+ HLH% intonational contour. *Paper presented at the Annual Meeting of the Linguistic Society of America*, New York, NY.

Watson, D. G., Arnold, J.E., and Tanenhaus, M.K. (2008). Tic Tac TOE: Effects of predictability and importance on acoustic prominence in language production. *Cognition, 106*, 1548–1557.

Wennerstrom, A., and Siegel, A.F. (2003). Keeping the floor in multiparty conversations: Intonation, syntax, and pause. *Discourse Processes, 36*, 77–107.

Wright, R. (2004). Factors of lexical competition in vowel articulation. In J. Local, R. Ogden, and R. Temple (eds.), *Phonetic interpretation: Papers in Laboratory Phonology VI*. (pp. 75–87), Cambridge: Cambridge University Press.

Xu, Y. (1999). Effects of tone and focus on the formation and alignment of F0 contours. *Journal of Phonetics*, 27, 55–105.

Yuan, J., Brenier, J.M., and Jurafsky, D. (2005). Pitch accent prediction: effects of genre and speaker. In *Proceedings of Interspeech 2005*, 1409–1412.

Appendix 1. Target items

The 48 critical sentences in the experiment are recoverable as follows. There are 12 conditions, formed by combining three types of question-response pairs (X-Z) and four kinds of object nouns in the responses (A-D). Each condition has four items, which can be differentiated based the verb-location context where the object nouns occurs (1–4). The subject of a question always consists of two personal names; no personal name occurs more than once in the experiment.

1. Context: *got…at the sports store*
   - (X) Narrow Corrective Focus
     Partner asks: *I heard that {Dawn and Alice; …} got gloves at the sports store.*
   - (Y) Narrow New-Information Focus
     Partner asks: *What did {Rachel and Carolyn; …} get at the sports store?*
   - (Z) VP/Wide Focus
     Partner asks: *What did {Angela and Joyce; …} do?*
   - (A) High Frequency and High Probability: *balls*
   - (B) Low Frequency and High Probability: *cleats*
   - (C) High Frequency and Low Probability: *fish*
   - (D) Low Frequency and Low Probability: *toys*

   Participant responds: ***(No,) they got {balls; cleats; fish; toys} at the sports store.***

2. Context: *kicked…in the garage*
   - (X) Narrow Corrective Focus
     Partner asks: *I heard that {Teresa and Martha; …} kicked dirt in the garage.*
   - (Y) Narrow New-Information Focus
     Partner asks: *What did {Connie and Sharon; …} kick in the garage?*

(Z) VP/Wide Focus
Partner asks: *What did {Evelyn and Jacqueline; ...} do?*
(A) High Frequency and High Probability: *cars*
(B) Low Frequency and High Probability: *cans*
(C) High Frequency and Low Probability: *books*
(D) Low Frequency and Low Probability: *shells*
Participant responds: ***(No,) they kicked {cars; cans; books; shells} in the garage.***

3. Context: *found...in the sea*
(X) Narrow Corrective Focus
Partner asks: *I heard that {Bonnie and Laura; ...} found boats in the sea.*
(Y) Narrow New-Information Focus
Partner asks: *What did {Mary and Irene; ...} find in the sea?*
(Z) VP/Wide Focus
Partner asks: *What did {Lillian and Gladys; ...} do?*
(A) High Frequency and High Probability: *fish*
(B) Low Frequency and High Probability: *shells*
(C) High Frequency and Low Probability: *balls*
(D) Low Frequency and Low Probability: *cans*
Participant responds: ***(No,) they found {fish; shells; balls; cans} in the sea.***

4. Context: *found...on the stairs*
(X) Narrow Corrective Focus
Partner asks: *I heard that {Matthew and Edward; ...} found socks on the stairs.*
(Y) Narrow New-Information Focus
Partner asks: *What did {Joseph and Steven; ...} find on the stairs?*
(Z) VP/Wide Focus
Partner asks: *What did {Daniel and Jason; ...} do?*
(A) High Frequency and High Probability: *books*
(B) Low Frequency and High Probability: *toys*
(C) High Frequency and Low Probability: *cars*
(D) Low Frequency and Low Probability: *cleats*
Participant responds: ***(No,) they found {books; toys; cars; cleats} on the stairs.***

Appendix 2. Lexical frequency of the target words

Word	Frequency in SUBTLEXus (per million)
fish	83.49
books	67.76
cars	45.63
balls	40.16
toys	13.22
cans	7.67
shells	5.57
cleats	0.41

Melanie Weirich

Friedrich-Schiller-Universität Jena

Organic Sources of Inter-Speaker Variability in Articulation: Insights from Twin Studies and Male and Female Speech

Abstract: This chapter presents three studies dealing with articulatory inter-speaker variability in German. In particular, organic sources (such as biomechanics of the tongue muscles, palatal shape and vocal tract dimensions) of idiosyncratic variation are discussed. Two studies deal with the within-pair similarity of identical (monozygotic) and non-identical (dizygotic) twin pairs; the third study describes differences between male and female speakers. The speech material comprises looping movements of the tongue in /aCV/-sequences, the production of the sibilant contrast /s/-/ʃ/ and the tense vowels /iː eː aː oː uː/ in different accent conditions. Results show that individual differences in articulatory strategies can at least in part be explained by idiosyncratic physiological restrictions and that the investigation of phonemic contrasts instead of targets, and the emphasis on speech dynamics are particularly relevant.

1. What we can learn from variability in speech

Research within the framework of speech perception has long dealt with the question of *invariance* in the speech signal. Possible invariant correlates of the speech production task in the physical space have been claimed to exist in various dimensions including articulation, acoustics and neural patterns (*Acoustic Invariance Theory,* Stevens and Blumstein, 1978; *Adaptive Variability Theory,* Lindblom 1988, 1990; *Motor Theory* Liberman et al., 1967; Liberman and Mattingly, 1985). However, we know that speech perception is multimodal and the articulatory movements and the acoustic signal are taken into account when both modalities are available (e.g. McGurk and MacDonald, 1976). In addition, numerous studies investigating intra- and inter-speaker variability – both in acoustic and articulatory terms – show that no true invariance exists and various combinations of physical correlates in both the acoustic and articulatory domain are present. Moreover, the variability found is not random, and should not be considered as distracting noise. Rather, we should consider it to be highly informative, telling us something about the speaker (or the respective speaker group), comprising both physiologically based restrictions as well as learned speech behavior (Foulkes and Docherty, 2006). In other words, a main question we are dealing with is, which variability is not just noise but systematic and is explainable due to which classifiable factors? Most generally, these factors can be separated into two potential sources, i. e. organic and learned (Ladefoged and Broadbent, 1957), and thus can be discussed within the *nature-nurture* framework. Of course, in most cases just one of these factors is never sufficient to fully explain the variability found, but it might be that sometimes one of the factors outweighs the other. The question is: When? And why? If we understand the reasons (when is which factor more important) we can learn something essential about the functioning of the speech production process.

The aim of studying variability is thus not to describe speaker-specific behavior per se but to determine particular groups of speakers that show the same "speaker-specific" behavior or strategies and to relate this variability with particular factors that classify the respective groups, arising from different biological, social or cultural sources. The studies described in this chapter focus on inter-speaker variability that is due to biological/organic variability. Biological similarity is present in related speakers, and in the most extreme way, in *twins*, the speaker group investigated in the first two studies of section 3. Biological similarity is also a strong factor when sex-specific differences are concerned and section 4 deals with articulatory differences between male and female speakers.

2. Learned vs. organic sources of inter-speaker variability

From psychological theories of learning, e.g., *Social Learning Theory* (Bandura, 1977) we know that people in general learn by observing and mimicking. Regarding language acquisition this implies that children learn the syntactic and prosodic structures, phonological patterns and lexical entries of a language through imitation of the people around them (i. e. especially in the beginning, mothers). Also dialectal pronunciation and sociolinguistic parameters of the parents are observed and absorbed by the child (Chambers, 2003). Moreover, this learning is a life-long process as has been shown very effectively in the analysis of pronunciation changes in the Christmas broadcasts of Queen Elisabeth II over a span of 40 years (Harrington, 2006). Sociolinguistic studies in general have shown that inter-speaker variation has numerous behavioral sources and can be used to create, express and attribute a certain social identity (for an overview see Foulkes and Docherty, 2006).

Nevertheless, organic sources of speaker-specific articulation exist which constrain the degrees of freedom a speaker has. Lindblom (1983, p. 217) assumes in his theory on the economy of speech gestures that "languages tend to evolve sound patterns that can be seen as adaptations to biological constraints of speech production." These biological constraints are manifold and comprise the length and constitution of the vocal folds, the size and dimensions of the vocal tract, the functioning of the tongue muscles, the shape of the palate and also the teeth. All of these organic factors can differ to some degree between speakers (or speaker groups such as adults and children or male and female speakers) and thus influence variation in articulatory strategies. Speaker groups (e.g. males vs. females, adults vs. children) differ in formant patterns due to biologically determined differences in the individual cross-sectional area of the vocal tract, with children showing the highest formant frequencies and males the lowest (Fant, 1960). Other studies have found a relationship between vocal tract geometry and articulatory space (Winkler et al., 2006; Fuchs et al., 2008). In particular, the individual articulatory distances between corner vowels (investigated using MRI in 9 French speakers) depended on the length of the speakers' pharynx: speakers with longer pharynxes showed larger degrees of freedom in the vertical direction and had larger vertical displacements than speakers with shorter pharynxes. Vocal tract size and dimension is the biological factor discussed in section 4 on sex-specific differences in articulatory spaces.

Several studies have emphasized the significant role of palate shape in articulatory variability (Lammert et al., 2013; Rudy and Yunusova, 2013; Brunner et al., 2009; Fuchs et al., 2006). For example, Brunner et al. (2009), found a relationship between a speaker's variability in tongue height and the steepness of the palate. Speakers with flat palates were more constrained in their variability than speakers with domed palates. The authors suggest that this is due to the large consequences on the area function/the acoustic output that small variation in tongue position can have in speakers with flat palates. Rudy and Yunusova (2013) showed that palate curvature and length can at least in part explain tongue position variability in the production of front consonants. They investigated VCV-sequences with C including stops, fricatives and affricates in 21 speakers of Canadian English. Lammert et al. (2013) investigated the interplay of hard palate morphology, articulation and acoustics in real vowel production data (MRI, five speakers) and in simulations. While simulations showed that palatal morphology affects formant frequencies, no significant correlation was found between real formant data and lingual articulation, leading the authors to conclude that speakers adapt their articulation strategies to accommodate palate shape differences. Palate shape as a potential organic source of inter-speaker articulatory variability is the factor investigated in the second twin study of section 3.

All three studies presented in this chapter concentrate on lingual inter-speaker variability that might be explained by organic sources, in particular, the palate shape (section 3, second twin study) and vocal tract dimensions (section 4, sex-specific differences in articulatory spaces). In addition, the first twin study of section 3 examines looping movements of the tongue during VCV-sequences in identical and non-identical twin pairs. By looking at the whole movement or gesture of the sequence, the influence of the tongue muscles, vocal tract dimensions and palate morphology is taken into account. It should be noted that in this chapter the term *gesture* is not used in the sense of an abstract idea (following Browman and Goldstein, 1992) but as a concrete movement of specific articulators.

3. Speaker-specific articulation in twins' speech

To investigate individual differences and to explain the variation in terms of the two possible influencing factors nature (i.e., genes and physiology) and nurture (i.e., environmental factors), a standard procedure in the field of behavioral genetic research is conducting twin studies (Spinath, 2005). Twin studies comprise a systematic comparison of the within-pair similarity of monozygotic (MZ) twins (who are 100% genetically identical) and dizygotic (DZ) twins (who share only about 50% of their genes, same as normal siblings). That anatomical and physiological characteristics are genetically determined and more similar in MZ twin pairs than in DZ twin pairs has been shown in several medical and dental studies, e.g. regarding the size and position of the jaw, the tooth size and the occlusal morphology (Lundström, 1948; Kabban et al., 2001) but also regarding the thyroid volumes (Langer et al., 1999). Eguchi et al. (2004) found in their comprehensive study of 78 male and female MZ and DZ twin pairs a high genetic contribution to speaker-specific variation in dental arch width, length, and also palatal height. While the twin types differ in their genetic/physiological similarity, they do not differ in terms of social environmental factors that contribute to the resemblance between individuals who grow up in the same family. If no particular emphasis by the parents is laid on treating the twins differently, they go to the same school, share most of their friends and also hobbies. In addition, for both twin types, the siblings have the same age at the same time, thereby being influenced by historical events in a similar way (*Equal Environments Assumption*, Scarr and Carter-Saltzman, 1979). Regarding language, both twin types share a) their environment during the speech acquisition process and b) social factors (such as school, hobbies and peer groups) which influence the speech of an individual. Thus, by comparing the within-pair similarity between MZ and DZ twins (who have grown up together, shared their speech acquisition process, and have a history that is not significant for differences in external factors such as surgeries, accidents, drug abuse or even the use of a pacifier that affects palatal shape during maturation), the role of physiological determinants and inherited morphological parameters can be analyzed. In other words, if MZ twins are more similar than DZ twins in a particular parameter, this parameter is affected by organic (genetic) factors.

While twin studies have a long tradition in the field of behavioral genetics research and go back to the late 19th century (Sir Frances Galton, 1876), analyses of twins' speech is rather new. Several studies have investigated speech acquisition and speech pathology in twins (Locke and Mather, 1989; Ooki, 2005; Simberg et al., 2009). However, only few have examined inter-twin variability in normal speech, and here, perceived similarity and acoustic features have been the predominant topics (see Loakes, 2006 and Weirich, 2012 for a more comprehensive overview of these studies). In summary, MZ twins have been found to be more similar than same-sex DZ twins or age-matched siblings in their average fundamental frequency

(Przybyla et al., 1992; Debruyne et al., 2002), voice quality parameters (van Lierde et al., 2005) and coarticulatory/dynamic patterns (Nolan and Oh, 1996; Whiteside and Rixon, 2003; Weirich, 2012). While perception studies have shown that familiar listeners can distinguish MZ twins (Whiteside and Rixon, 2000), unfamiliar listeners succeed in distinguishing unrelated speakers by using only one short bi-syllabic word but fail to do so in both MZ and DZ twin pairs (Weirich and Lancia, 2011).

Articulatory studies in twins have rather been neglected (but see Weirich, 2012). A reason for the uncommonness of articulatory studies in twins might be a methodological one: articulatory analyses in general involve only small subject groups due to their time-consuming character, and together with the fact that in twin studies we usually compare the similarity of speaker pairs (i. e., MZ twin pairs vs. DZ twin pairs), the problem of a small subject group becomes even more crucial. Moreover, the participating twins have to fulfill several requirements concerning environmental factors such as the time they have spent (and are still spending) together, surgical interventions and also the attitude they have towards being a twin (a negative attitude could lead to an enhancement of individuality also expressed in an idiosyncratic speech style).

However, despite these difficulties, twin design studies have a high potential for helping us to distinguish physiological determinants and environmental factors responsible for individual differences in speech. The impact of physiological factors is especially relevant with regard to speaker-specific articulation strategies. Thus, if we are interested in understanding the reasons for inter-speaker variability, analyzing articulatory variability in MZ and DZ twin pairs is a promising source of information. Therefore, one of the two main aspects of this chapter is the discussion of two recent studies that we have conducted on the speech of MZ and DZ twins, concentrating on within-pair articulatory variability in VCV sequences (Weirich et al., 2013) and in the realization of the sibilant contrast /s/-/ʃ/ (Weirich and Fuchs, 2013).

3.1. Individual articulatory strategies in looping movements

The first study presented here on articulatory variability in twins' speech is on a particularly interesting articulatory gesture: the looping movement of the tongue (for a more detailed description of the study see Weirich et al., 2013). Loops are curved trajectories of the tongue back found in VCV-sequences where C is a velar consonant. The trajectories of the sequence do not simply consist of straight lines between vowel and consonant targets but an elliptical movement of the tongue back – a loop – is found (Kent and Moll, 1972, Mooshammer et al., 1995; Hoole et al., 1998; Löfqvist and Gracco, 2002; Geng et al., 2003; Perrier et al., 2003; Brunner et al., 2011). Curved paths in movements in general have been shown to be potentially explained by anatomical factors and muscle mechanics (see Flanagan et al., 1993; Gribble and Ostry, 1996; Gribble et al., 1998 for arm movement, Perrier et al., 2003; Perrier and Fuchs, 2008 for orofacial movements). Thus, it is argued that the loops of the tongue are also a result of the biomechanical characteristics of the muscles and the surrounding vowel targets (Perrier et al., 2003). From that we can hypothesize that loops should be more similar in MZ twins than in DZ twins or unrelated speakers. If, on the other hand, loops are actively controlled (Löfqvist and Gracco, 2002) and reflect learned behavior independent of individual physiology, the degree of variability within a twin pair should depend less on the twin type (MZ vs. DZ).

3.1.1. Articulatory analysis

3.1.1.1. Participants

The participants were ten German speakers (20–34 years old): two female DZ twin pairs and two female and one male MZ twin pair. All speakers were born, raised and still living in Berlin, Germany. The twins grew up together and were still seeing each other at least twice a month. With a comprehensive questionnaire we controlled for differences in potential influencing factors such as relevant surgeries, habits or attitudes towards being a twin, but also the behavior of the parents in raising their children was checked (e.g. treating them particularly different or making them like the same things). All of the participants liked being a twin and no pair differed with respect to surgeries, accidents or habits (e.g. use of pacifier as a child, singing, smoking) that might have affected physiological characteristics of the speech apparatus. Also, all twins reported being treated similarly by the parents and having shared friends and hobbies especially during childhood and adolescence.

In addition to the comparison of speakers within the same twin pair (groups MZ and DZ), speakers of different twin pairs were paired to form the group of unrelated (sex-matched) speakers (group UN).

3.1.1.2 Recordings, speech material and measurements

Acoustic and articulatory recordings were conducted in the speech lab of ZAS (Zentrum für Allgemeine Sprachwissenschaft, Berlin) by means of 2D electromagnetic articulography (EMA, Carstens AG100). Two coils, one above the upper incisors and one at the bridge of the nose served as reference coils and were used for head movement correction. Three coils were attached to the tongue (one approximately 0.5 cm behind the tongue tip, one approximately 5 cm behind the tip on the tongue back, and a third one halfway in-between the two, on the tongue dorsum). Comparable positioning of the coils within the twin pairs was attained using a true-to-scale template of the tongue with the coils of one of the twins (created with the help of a printed photograph) being used as a reference for the second twin. For the analysis of the looping movement, we concentrated on the coil positioned furthest back on the tongue (henceforth, tongue back coil).

The speech material was obtained during a larger recording session with different target phonemes and carrier sentences (see Weirich, 2012). For the analysis of the looping pattern of the tongue back the sequence /aCV/ within the names "Haga", "Hagu", "Haka", and "Haku" was chosen. The target words were part of the sentence "Ich grüße/wasche Haka/Haga/Haku/Hagu im Garten" (I greet/wash Haka/Haga/Haku/Hagu in the garden). On average 9.45 repetitions for each speaker and each /aCV/ sequence could be used for the analysis.

For a comparison of the shape of the looping patterns between speakers, including all repetitions, the data had to be processed in several ways. Briefly, first, the shape of the looping movement had to be parameterized. Therefore, instead of taking absolute positions of the tongue back coil (in vertical and horizontal dimension), curvature was calculated for each measurement point throughout the /aCV/-sequence (cf. Tasko and Westbury, 2004; O'Neill, 2006). This was also done to prevent a confound of the potentially more similar coil positions in MZ twins than in DZ twins or unrelated speaker pairs. Second, multiple pairwise comparisons were done (separately for each /aCV/ sequence) consisting of either trajectories from two speakers of the same twin pair (MZ or DZ) or trajectories from two unrelated (sex matched) speakers (e. g. twin 1 from twin pair A and twin 1 from twin pair B). Third, the articulatory trajectories had to be temporally normalized. For this reason we adopted the functional data analysis tool proposed by Lucero et al. (1997) and used a registration method described in detail in Lancia and Tiede (2012) to get time-aligned trajectories for each /aCV/-sequence and each speaker pair separately. Fourth, distances were measured between all points of each pair of aligned curvature data and mean distances were calculated for each comparison to be used for the statistical analysis. For further details on the recording session, the labeling procedure and the different processing steps see Weirich et al. (2013).

3.1.2. The impact of physiology on looping movements

For purposes of exemplification Figure 1 shows looping trajectories during /aka/ for two speakers (in grey and black) of a female MZ twin pair (left) and two speakers (grey and black) of a female DZ twin pair (right). The plots show the positional data (in horizontal and vertical dimensions) of the average tongue movement (without time-alignment), the arrow marks the direction of movement. The MZ twins in the left plot exhibit similar shapes of their loops, starting with a rather straight/slightly curved upwards movement from the vowel to the velar stop, a horizontal – possibly sliding – movement along the palate and a straight or slightly backwards oriented downwards movement to the second /a/. The loops of the DZ twins in the right plot show more obvious differences in shape and looping characteristics. While speaker DZa (grey) resembles the MZ speakers in terms of a slightly forward directed upwards lifting, horizontal movement (even though to a lesser extent) and a downward and backward movement to the second /a/, DZb (black) shows a more s-like shape of the tongue lifting, no sliding along the palate with a very steep angle at the turning point and a backwards directed movement to the second vowel. This pair also differs with respect to the relative positions of the two vowels: while DZa (grey) produces V2 at a more fronted position than V1 (as the two speakers of the MZ pair), speaker DZb (black) produces V2 at a more retracted position than V1. Note that for the statistical analysis, curvature and not positional data was used. A high value in curvature corresponds to a change in movement direction, while low curvature values reflect rather straight movements. Thus, DZb is the speaker revealing the clearest peak in curvature over the whole movement.

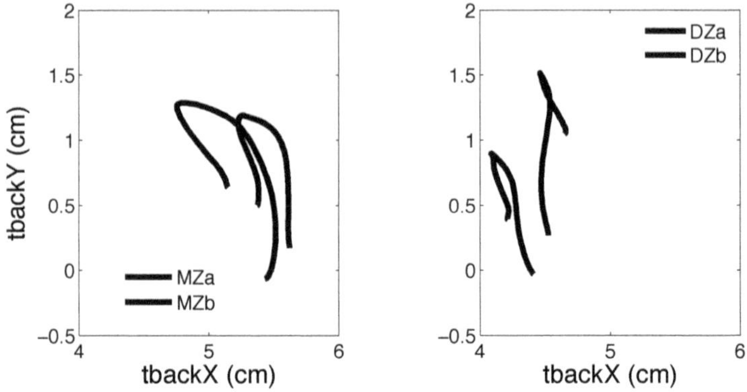

Figure 1: Mean looping trajectories of the tongue back coil during /aka/ for two twin pairs (left MZ, right DZ). Different speakers of each pair are indicated by grey and black. Vertical movement of the tongue back is displayed on the y-axis, horizontal movement on the x-axis (in cm). Reprinted from Journal of Acoustical Society of America *134, 5, 3766–3780. Weirich, M., Lancia, L., Brunner, J. Interspeaker articulatory variability during vowel-consonant-vowel sequences in twins and unrelated speakers. Reproduced with permission from AIP Publishing LLC. Copyright 2013.*

For the statistical analysis a linear mixed model (Pinheiro and Bates, 2000) as implemented in the lme4 package of the R software (version 2.14.1, R Development Core Team, 2008) was run. The dependent variable was the measured pairwise mean absolute distance between the aligned curvature data. The logarithmic values of these distances were used to normalize the residuals, a mandatory assumption in linear mixed models (Pinheiro and Bates, 2000). We included speaker group (with the levels MZ, DZ and UN = unrelated speakers), vowel (/a/ vs. /u/) and voice (voiced vs. voiceless) as fixed factors and a pair specific random intercept for vowel and voice.

Figure 2 shows the distribution of the log transformed distance measures for all /aCV/-sequences together but separated by speaker group. As is apparent in the figure, the statistical analysis revealed a significant difference between MZ twins and DZ twins (pMCMC < 0.001), but not between DZ twins and unrelated speaker pairs. No interaction between vowel and speaker group or voice and speaker group was found but a three way interaction between all factors suggested that for the vowel /a/ a stronger effect of the comparison between MZ and DZ twins exists in the voiced than in the voiceless condition.

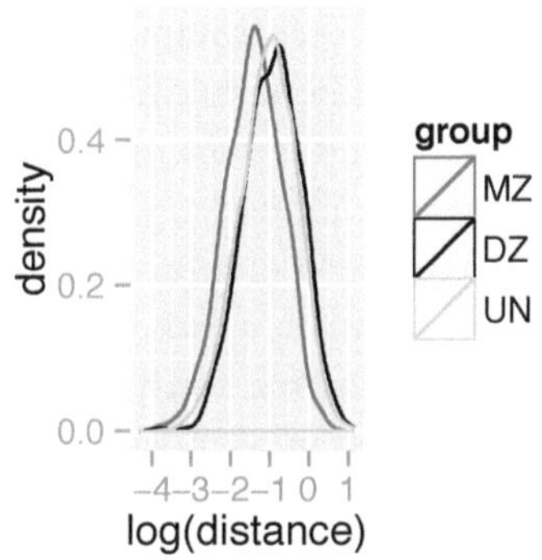

Figure 2: Distribution of logarithmic distances in curvature for all /aCV/-sequences separated by the three groups: monozygotic twins (MZ), dizygotic twins (DZ) and unrelated speakers (UN).

3.1.3. Discussion

The results of the study reveal a significant influence of shared physiology on articulatory inter-speaker variability. In detail, more similar looping patterns in VCV sequences were found for MZ than for DZ twins or unrelated speakers. By investigating articulatory movements (such as loops) and not articulatory target positions the focus has moved from static to dynamic aspects of the speech signal. Particularly regarding inter-speaker variability this might be an essential factor. Nolan et al. (2006) suggested that the speech signal can be described by two different aspects: 1) linguistically determined targets and 2) organically determined transitions. It is proposed that while the targets are constrained by the shared language system and carry the linguistic information, the transitions link the adjacent targets and are more prone to reflect speaker-specific characteristics that are due to individual physiology.

Recently, studies on inter-speaker variability have focused not only on phonemic targets (or transitions) but also on the realization of phonemic *contrasts*. In this way, the phonetic inventory of a language is better reflected and taken into account. The next twin study deals with the realization of the sibilant contrast /s/-/ʃ/ in German.

3.2. Individual articulatory strategies in realizing the sibilant contrast

Toda (2006) found two different speaker-dependent strategies in the realization of the sibilant contrast /s/ - /ʃ/ in French: 1) tongue placement strategy (where speakers only retract their tongue horizontally) and 2) tongue adjustment strategy (where speakers additionally elevate their tongue). With respect to sibilants, some of the most important work in recent years has been conducted by Perkell and colleagues (Perkell, 2010; Perkell et al., 2004). Their work particularly emphasizes the link between speech production and perception. In other words, they find that speakers with poorer auditory acuity of a phonemic contrast also tend to produce this contrast less distinctively. Ghosh et al. (2010) went one step further by including a speaker's somatosensory acuity (which implies the sensation of touch, i. e. tactile feedback) into the analysis of the acoustic realization of /s/ and /ʃ/. They found a positive correlation between a speaker's acoustic distance between the sibilants and their auditory and somatosensory acuities. When tactile feedback plays a role in the realization of a phonemic contrast, as Ghosh and colleagues found, then individual differences in the respective morphological structures relevant for the sound production (i. e. the palatal shape) might also affect the realization of this contrast. Perkell et al. (2004) included some morphological parameters in their analysis of the sibilant contrast. They examined palatal height, length and width but could not find any significant correlations. They did, however, not include a parameter that is essential for the production of sibilants: the palatal and in particular the alveolo-palatal steepness. Thus, in our study (Weirich and Fuchs, 2013) we investigated the potential relationship between speaker-specific realizations of the /s/-/ʃ/ contrast in German and the speaker's palatal shape, parameterized by two angles describing the overall steepness of the palate and the steepness of the alveolo-palatal ridge, where the contrast is realized.

3.2.1. Articulatory analysis

3.2.1.1. Participants

The study consisted of two different experiments (EMA and EPG) with different speaker samples. The EMA study comprised the same DZ and MZ pairs from the VCV study (4 female and 1 male pair). In addition, another male MZ pair, part of the twin corpus recorded at the ZAS (Weirich, 2012) could be included. The EPG experiment comprised 12 unrelated German speakers (7 females and 5 males) with no hearing or speech impairments, aged between 24 and 56.

3.2.1.2. Recordings, speech material and measurements

The speech material of the EMA experiment was acquired during the larger recording session of the Weirich (2012) study. The target sounds were the sibilants /s/ and /ʃ/ that were part of the German verbs /kʏsə/ (1. p. sg. of 'to kiss') and /vaʃə/ (1. p. sg. of 'to wash') embedded in carrier sentences. On average, 32 repetitions for each speaker and phoneme were included. The target positions of /s/ and /ʃ/ were labeled oriented on the minimal tangential velocity of the tongue tip sensor. We then investigated inter-speaker variability in realizing the contrast in terms of the horizontal and vertical position of the tongue tip following Toda's (2006) idea of two different speaker specific strategies which vary in the amount of vertical tongue elevation. While we cannot compare the whole overall shape of the tongue, as Toda did, due to the use of EMA-data which gives us information only about the position of three flesh points on the tongue, we can compare the position of the tongue tip between the two sibilants for each speaker and thereby investigate the vertical/horizontal distance between the sound productions.

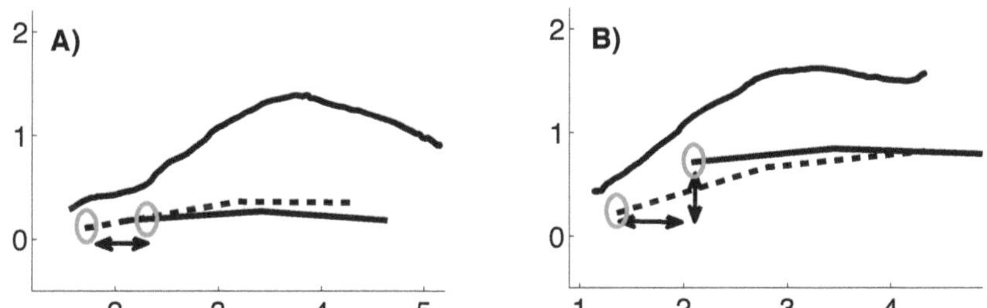

Figure 3: Distance measurement (in horizontal and vertical dimensions) between mean tongue tip positions (dashed line = /s/, solid line = /ʃ/) for two speakers of different twin pairs. Reprinted from Journal of Speech, Language, and Hearing Research *56, 8, 1894–1190, Weirich, M. and Fuchs, S. Palatal morphology can influence speaker-specific realizations of phonemic contrasts. Reproduced with permission from the American Speech-Language-Hearing Association (http://jslhr.pubs.asha.org). Copyright 2013.*

Figure 3 shows two strategies in two of our participants and visualizes their mean interpolated tongue contours during their articulatory target positions for /s/ (black line) and /ʃ/ (dashed line): while speaker A only retracts the tongue for /ʃ/ in contrast to /s/, speaker B retracts and additionally elevates the tongue, following the palate contour. To quantify this, for each speaker the horizontal and vertical distances between the tongue tip positions of the two sounds were summed up to 100%. The horizontal and vertical distance was then expressed in percentages, too, in relation to the total amount.

For the EPG experiment, the sibilants /s/ and /ʃ/ were recorded on average 30 times per speaker, they occurred in the nonsense words /zasa/ and /ʃaʃa/ and were embedded in a carrier sentence. We defined the place of articulation for the two sounds using the articulatory center of gravity (COG, Hardcastle et al., 1991), which is a weighted index that attaches more importance to rows at the front of the palate. A higher COG thus reflects a more anterior place of articulation (typical for /s/). The differences in COG between the /s/ and /ʃ/ productions of each speaker were then calculated to get a distance measure comparable to the one used in the EMA study.

Physiological measures were taken regarding body size, body weight and tongue (for the twin study) and the palate (for both studies). The different measurements were taken to look for their potential impact on sibilant production but also to confirm the assumption that physiological parameters were more similar in the MZ than in the DZ twins (for further information, see Weirich and Fuchs, 2013). Here, we will concentrate on the most crucial parameter in sibilant production: the palatal shape. The palatal shape was parameterized by two different angles: the angle of the overall palatal steepness (δ), and the angle of the alveolo-palatal ridge (γ) shown in Figure 4.

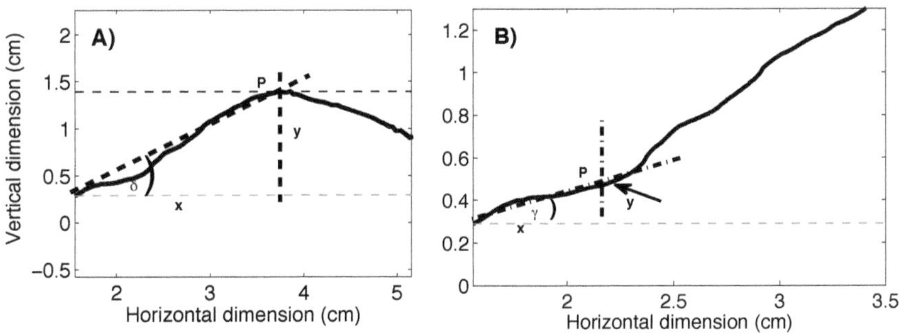

Figure 4: Visualization of angle measurements: angle of general palatal steepness δ (A) and angle of alveolo-patalal ridge γ (B, close up view). The thick black line shows the palate contour, the thinner dashed horizontal lines the minimal and maximal vertical positions of the palate. P defines the highest point of the palate (in A) or the alveolar step (in B, see arrow) with corresponding vertical (y) and horizontal (x) interval. Reprinted from Journal of Speech, Language, and Hearing Research *56, 8, 1894–1190, Weirich, M. and Fuchs, S. Palatal morphology can influence speaker-specific realizations of phonemic contrasts. Reproduced with permission from the American Speech-Language-Hearing Association (http://jslhr.pubs.asha.org). Copyright 2013.*

The calculation of these angles was done in the same way for both experiments and is expressed in equation (1):

$$\tan(\delta, \gamma) = y(P) / x(P), \qquad (1)$$

where P is the point on the palate that determines the height, y, and the length, x, necessary to calculate the particular angle. P differs for the two angles and reflects either the maximal vertical point on the contour for the angle δ (see plot A in Figure 4) or the visually defined position of the alveolo-palatal ridge for the angle γ (in most cases easily identifiable by a small dip as seen in plot B of Figure 4).

To look for a potential relationship between morphology and articulation, correlations were run between the two palate angles and the horizontal distance between the sibilants (in % for the EMA study or expressed as COG difference for the EPG study).

3.2.2. *The impact of palatal shape on articulatory realization of sibilant contrast*

The first main result of the EMA twin study was that we found more similar articulatory strategies in MZ twins than in DZ twins. This reflects the findings of the looping study shown above. While no difference between any of the four MZ twins was found (Welch two sample t-tests), both DZ pairs revealed significant differences ($p < 0.01$) in terms of their horizontal tongue tip variation (in %) between the two sounds.

The second main result was that we found a clear effect of individual palatal shapes on the articulatory realization of the sibilant contrast. For the twin study both angles revealed a significant negative correlation (Spearman) with the horizontal distance measure (in %): while the overall palatal steepness angle δ showed a correlation of –0.53 ($p < 0.05$), the correlation was even higher for the alveolo-palatal angle γ (–0.78, $p < 0.01$). The relationship of the latter angle to the articulatory realization is shown in Figure 5 (left plot). The smaller the angle (the flatter the palate) and the more horizontal distance (in %) is found. The figure also reveals the more similar articulation for the MZ twins (marked by the filled symbols) than the DZ twins (unfilled symbols).

Figure 5: Relationship between alveolo-palatal angle (x-axis) and articulatory realization of the sibilant contrast (y-axis). The plot on left side shows the EMA-twin-data, the plot on the right side the EPG-data (m: male speaker, f: female speaker). The black line shows the regression line and the gray shadowed area defines the 95% confidence interval.

The remaining question then was whether we can find this influence of palatal shape also in a more heterogeneous group of unrelated speakers. Figure 5 (right plot) shows the COG distance measure of the EPG study in dependence of the alveolo-palatal angle. Despite one outlier (a male speaker with an extremely high COG distance measure) a significant negative correlation of –0.62 ($p < 0.05$) was found, mirroring the results of the twin data.

3.2.3. Discussion

The study revealed that not only does individual physiology play a role in dynamic aspects of articulation (such as loops), but also in the realization of a phoneme contrast. The articulatory organization of a speaker's targets is affected by his/her speaker-specific organically determined idiosyncrasies. Especially in sibilants, where the tongue-palate contact is crucial, these individual physiological characteristics come to the fore and show their impact. In detail, the shape of the alveolar ridge – which is the articulatory place where the sibilants are produced – can account for at least some of the inter-speaker variability found in the articulation of sibilants.

The question arises whether other phonemes that are less affected by physiological restrictions such as vowels might also be affected by individual differences in vocal tract anatomy. The final study focuses on inter-speaker variability in articulatory vowel spaces. Here, the speaker groups under investigation are male and female speakers, which have been found to differ in the physiological characteristics essential for the production of vowels (such as the overall vocal tract size, and the relationship between oral and pharyngeal cavity dimensions).

4. Speaker-specific articulation in male and female speech

Most studies on differences between male and female speech have concentrated on acoustic differences, fewer have investigated potential articulatory variability. A very salient and highly investigated aspect is the larger acoustic vowel space in females. It has been found for several languages, such as American English (Diehl et al., 1996), British English (Whiteside, 2001), German (Weirich and Simpson, 2014a) and Swedish (Simpson and Ericsdotter, 2007). The differences between vowel spaces are not uniform, with differences between different vowel categories increasing as F1 and

F2 increase. Thus, male and female speakers differ most in front and low vowels (such as /i:/ and /a/) and less in high back vowels (such as /u:/) (Fant 1966). Various hypotheses have been proposed to account for this variability. While some focus on purely behavioral reasons, such as the sociophonetic explanation of females aiming at speaking more clearly than males (Bladon et al., 1983; Henton, 1995), others emphasize physiological (sex-related) differences. One of the latter is the non-uniform difference between males and females in the relationship of pharyngeal and oral cavity (Chiba and Kajiyama, 1941; Fant, 1966, 1975; Nordström, 1977; Winkler et al., 2006; Fuchs et al., 2008).

A third strand of possible explanations is based on acoustic-perceptual compensation (Goldstein, 1980; Ryalls and Lieberman, 1982; Diehl et al., 1996). The reasoning is as follows: The higher the fundamental frequency, the sparser the harmonics. The greater inter-harmonic spacing in higher pitched voices causes a poorer definition of the spectral envelope (and in particular of the formants). From that it is hypothesized that the larger acoustic distance between female vowel targets compensates for the poorer spectral definition more typically found in high-pitched female voices. However, in a recent study of 56 female speakers with varying fundamental frequency (from 154 Hz to 234 Hz), we did not find a correlation between f0 and acoustic vowel space size (Weirich and Simpson, 2013) suggesting other factors (organic and/or learned) must be responsible for the larger female acoustic vowel space.

Another explanation involves the underlying articulatory dynamics in producing the vowel space. Despite females having, on average, larger acoustic vowel spaces than males, Simpson (2001, 2002) found smaller articulatory vowel spaces in females than in males. In addition, Simpson (1998) found sex-specific differences in the relationship between formant values and duration (some expected correlations were only found for males but not females). Due to females exhibiting on average smaller vocal tracts than males they reach their articulatory targets earlier (in terms of time and space), and thus, might undershoot their targets less than males. Vowel undershoot can result from different degrees of coarticulation possibly induced by varying accent and stress conditions. Lindblom (1983, 1990) suggested in his H&H theory that speech varies along a continuum between output-oriented, hyperarticulated stressed syllables at one end and system-oriented, reduced/hypoarticulated unstressed syllables at the other end. Since then the relationship between stressed and unstressed syllables and hyper- and hypoarticulation has been investigated intensively (e.g. de Jong et al., 1993; de Jong, 1995, 1998; Harrington et al., 2000; Cho, 2004). Mooshammer and Geng (2008) investigated articulatory manifestations of vowel reductions in German and found a greater degree of coarticulation with the consonant context in unstressed vowels than in stressed vowels. If females reach their articulatory targets earlier/more often than males (e.g. even in unstressed vowels), then they should be less influenced by accent-induced undershoot. If no differences in undershoot were found between the sexes, we would expect to find higher velocities or longer durations in males, but this is not the focus of the present investigation.

To test this assumption we conducted an articulatory analysis of 4 female and 5 male German speakers including speech material suitable to investigate a speaker's "extreme" articulatory vowel space only minimally affected by coarticulation and accent-induced undershoot. The aim was then to use this as a speaker-specific articulatory reference frame that all further analyses could be compared to (Weirich and Simpson, 2014b).

4.1. Articulatory analysis

4.1.1. Participants and recordings

Five male and four female German speakers took part in the study. The speakers were between 23 and 43 years old and came from the Eastern Central German dialect area but showing very little dialectal influence. Articulatory recordings were made at Potsdam University with the NDI-Wave system. Parallel to the twin studies, three coils were attached to the tongue and, for the present analysis, the movement of the coil positioned furthest back on the tongue (tongue back coil) was investigated. The articulatory labeling was done with the help of the MATLAB based software *mview* developed by Mark Tiede (Haskins Laboratories).

4.1.2. Speech material

The speech material was part of a larger corpus comprising 20 different target words (approximately 10 repetitions each) in different accent conditions and varying carrier sentences. The data presented here is twofold: The first set of data included the three corner vowels /a: u: i:/ contained in the double vowel sequences in the abbreviation *IAA*, *AUU* and *BII*. The abbreviations were used because here the articulatory positions of the vowels were expected to be extreme and only minimally effected by coarticulatory influences. The second set of data included the sequence /gV/ with V being /i: e: a: o: u:/ in the German name *GVbi* embedded in the carrier sentence *Ich sah GVbi an* ('I looked at *GVbi*'). Here, three different accent conditions were recorded. First, the participants were asked to read the sentences presented to them from a screen (control condition, c). Second, participants produced the sentences in response to questions from the experimenter eliciting an answer with either the name under focus (accented condition, a) or the preceding verb (unaccented condition, u).

4.1.3. Sex-specific 'extreme' vowel spaces in IAU-polygon

Figure 6 shows the articulatory positions of the tongue back coil at the vowel targets /i: a: u:/ measured at the midpoint of the double vowel sequences of the abbreviations. The data was translated for each speaker with the midpoint of the vowel space set to the origin (0/0). This facilitated a better visual comparison between the sexes. The displayed data includes all repetitions of all male (black) and female (grey) speakers. As we can see, there is a tendency for male speakers to exhibit larger articulatory spaces than females (on average 93 mm^2 vs. 66 mm^2), and this variability is due to a vowel-specific difference: while the articulatory positions completely overlap for /u:/, males exhibit a lower and more retracted tongue position for /a/ and a higher tongue position for /i:/. Thus, statistical tests revealed a significant difference between males and females only for the mean Euclidean distance (ED) between /i:/ and /a:/ (Welch two-sample t-tests, $t = -2.7$, $df = 5.9$, $p < 0.05$). This is also expressed in the sex-specific dimensions of the space: while males on average exhibit a 1.3 times larger vertical than horizontal expansion, this relationship is around 1 for the females.

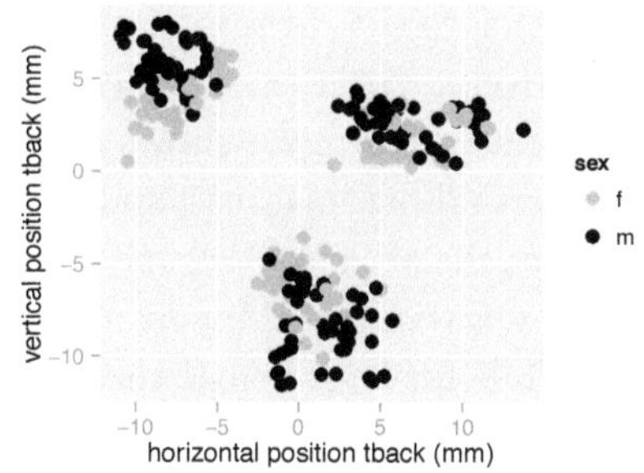

Figure 6: IAU-polygon: articulatory positions of the tongue back coil measured in the double vowel sequences of the abbreviation. Female speakers are in grey, male speakers in black.

4.1.4. Sex-specific differences in undershoot

Analyses of /gV/-sequences served two aims. The first one was to compare the vowel space resulting from the tense vowels produced in this sequence with the speaker-specific "extreme" reference vowel space resulting from the IAU-polygon. This made it possible to analyze the degree of coarticulation-induced undershoot individually for each speaker and then compare it between speakers, and ultimately sexes. The second aim was to compare the degree of accent-induced undershoot between males and females by analyzing the vowels of the /gV/-sequence in three different accent conditions (control, accented, unaccented).

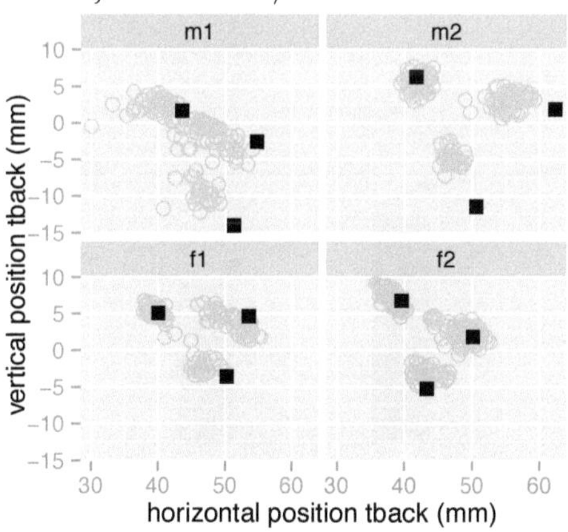

Figure 7: Articulatory positions of the tongue back coil during the IAU (extreme) vowel space (black squares) and the /gV/-sequence (grey circles) of two male (m1, m2) and two female (f1, f2) speakers.

Figure 7 gives a first hint of speaker- (or sex-) specific differences in the relationship between the "extreme" IAU vowel space and the coarticulatorily more affected /gV/-vowel space. The figure shows four subplots, visualizing the data of two male speakers (above, m1, m2), and two female speakers (below, f1, f2). For each speaker, the black squares show the mean IAU-polygon, the grey circles show the vowel space resulting from the /gV/-sequence for all three accent conditions. While for both female speakers the vowel spaces of the IAU and the /gV/-sequence overlap considerably, differences are apparent for both male speakers, especially in terms of a lower and more retracted position for /a/ in the IAU space compared to the /gV/-space.

The bars of Figure 8 show the average female (grey) and male (black) vowel space of the /gV/-sequence in absolute terms (in mm²) separately for the three accent conditions. The male speakers reveal higher values for all accent conditions; however, the difference is only considerable for the accented condition. In addition, the figure shows the average female and male vowel space of the /gV/-sequence in percent of the IAU-space (black and grey circles connected by the lines). The relationship between the vowel space produced within the /gV/-sequence and the "extreme" IAU vowel space was calculated for each speaker and accent condition separately. It is apparent that here, females reveal substantially higher values than males in the control and unaccented condition, while in the accented condition the considerable difference between males and females found for the absolute values is absent.

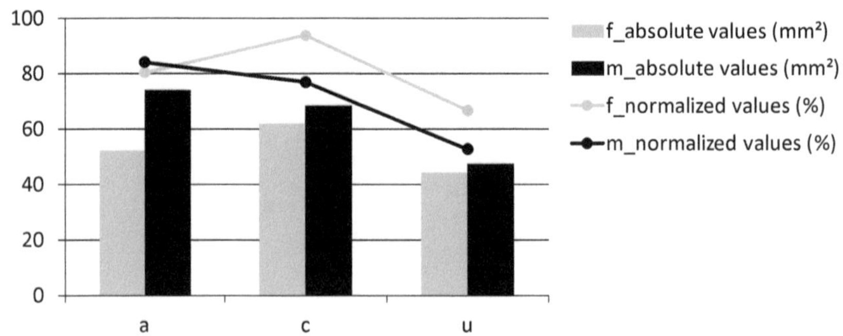

Figure 8: Average polygon sizes of the /gV/-data for male (black) and female (grey) speakers separated by accent condition (a, c, u). The bars represent the vowel spaces in absolute terms (mm²), the connected dots represent the vowel spaces in normalized terms (%).

For statistical analysis (linear mixed models), not the overall vowel space size per speaker, but rather EDs from the midpoint of the vowel space to each vowel were used as dependent variable. In this way, the number of data points could be increased and a vowel specific analysis could be undertaken. Two analyses were run, with either the absolute EDs as dependent variable or the ED expressed as a percentage of the EDs between vowels and the midpoint measured in the IAU data. Model comparisons (likelihood ratio tests) were conducted to find the model with the best fit to the data. For the absolute ED as dependent variable, we found a significant interaction of sex*vowel and sex*accent condition (random factors included were speaker and repetition). Regarding the first interaction, a significant difference between males and females was only found for the vowel /a:/ analogous to the results of the IAU polygon (estimate: 2.3, pMCMC-value < 0.01). Regarding the second interaction, males show a significant difference between accented and unaccented (estimate: –1.5, pMCMC < 0.01), while females do not.

For the normalized EDs as dependent variable, we found a significant interaction of sex*accent. In contrast to the absolute values, no sex-specific differences were found for the ED for /a:/ (or any other vowel). However, analogous to the absolute values, the factor accent condition showed its significance in terms of sex-specific differences: males differed between accented and unaccented condition (estimate: –17.6, pMCMC < 0.01), while females did not.

4.2. Discussion

Our results are in line with the hypothesized higher probability of accent-induced undershoot in males than in females: while males show the expected significantly smaller articulatory vowel spaces in unaccented conditions (in absolute and normalized values), females do not differ between the accent conditions. Additionally, the expected larger articulatory spaces in males were only found in the IAU-data, where articulatory positions are assumed to be "extreme" in terms of being minimally affected by coarticulation-induced undershoot. While we cannot rule out that females "do more" than males on purpose (in terms of reaching their articulatory targets irrespective of accent condition and coarticulatory influences to achieve a large acoustic vowel space resulting in clear speech), we have indeed seen that sex-specific differences vary depending on whether we look at the absolute or relative values: while males exhibit larger articulatory distances in absolute values, the difference is leveled out or even reversed in normalized values. This leads us to suggest that the same articulatory distance results in different acoustic outputs in speakers who differ in their maximal articulatory spaces (such as an average male and an average female speaker). To examine this more closely we are currently investigating potential sex-specific differences in the articulatory-acoustic relationship in the production of diphthongs.

Thus, sex-specific differences in acoustic vowel spaces might be due to differences in anatomically restricted articulatory spaces between males and females. We suggest that the underlying dynamics of the articulatory gestures play a crucial role in sex-specific differences.

5. General discussion

Various sources of inter-speaker variability, including behavioral and organic factors exist and all of them are worthy of systematic investigation and categorization. In this chapter the impact of organic factors on inter-speaker variability was highlighted by investigating two speaker groups in which biological variation is a central issue: related speakers (twins) and male and female speakers. We have seen that individual differences in lingual strategies can at least in part be explained by idiosyncratic physiological restrictions. In particular, the shape of the palate, the physiological properties of the tongue muscles and the size and shape of the vocal tract seem to be crucial factors regarding articulatory inter-speaker variability. The variability found is systematic and explainable, and can help us understand some of the underlying principles of the speech production process. Following Lindblom's assumption that languages "tend to evolve sound patterns that can be seen as adaptations to biological constraints of speech production" (1983, p. 217), it is suggested, that also at least some of the inter-speaker variability we have discussed mirrors speaker-specific adaptations to individual biological restrictions (see also Lammert et al., 2013). While in many phonological theories speaker-specific behavior is considered a source of random noise with no impact on phonemic categories, we found a significant influence of individual differences in the alveolo-palatal steepness on inter-speaker variability in realizing the phonemic contrast of /s/-/ʃ/ in German and thus, on the phonetic realization of two phonemic categories.

Furthermore, in addition to investigating phonemic contrasts instead of targets, another crucial step in the analysis of inter-speaker variability is to focus more on the dynamic aspect of speech. In line with Nolan et al. (2006), who suggested the speech signal contains linguistically determined targets and organically determined transitions, we found a significant influence of organically determined differences on the looping movements of the tongue (i. e. transitions) during /aCV/-sequences. While it is recognized that the movement is of course also affected by the targets (and here especially the stop closure at the palate), we suggest that dynamic patterns in speech are especially appropriate for showing the influence of organic sources on inter-speaker articulatory variability. Whether the properties of the tongue muscles (biomechanics) or the palate shape/vocal tract dimensions (physical constraints) are the chief influencing factor remains to be examined.

Another way of highlighting the underlying dynamic nature of articulatory gestures is to set the lingual movement in relation to the size and shape of a speaker's individual and organically determined articulatory space, instead of comparing the absolute sizes of the movement. The same articulatory movement (in shape and size) might be extreme for a small female speaker but only half of the potential movement size of a large male speaker. If there is enough time for the gesture (as e.g. in an accented position) both speakers will reach their extreme target positions. These maximal positions will differ according to the speakers' respective physiological space, as was shown for the IAU vowel spaces. If time is short (for example due to a target's occurrence in an unaccented position) the female speaker might reach her extreme position, while the male speaker reaches only 50% of the movement amplitude he could reach when there were enough time. This might be one reason for the sex-specific differences we found in accent-induced undershoot: while males revealed significantly smaller amplitudes in unaccented conditions than in accented conditions, females did not differ. Going a step further, it could be suggested that the same articulatory gesture (in shape and size) results in different acoustic outputs. This is especially interesting in the light of the mismatch between articulation and acoustics regarding sex-specific differences in vowel space sizes: despite having smaller articulatory vowel spaces, females exhibit larger acoustic vowel spaces than males (Simpson 2002). In our current work we are investigating the acoustic vowel spaces of the IAU-data, and while males showed larger articulatory distances, the acoustic distances between the vowels did not differ between the sexes. In addition, we are examining acoustic and articulatory diphthong realizations in males and females and while no significant sex-specific difference in absolute articulation is found, males and females differ in their respective acoustic output. In both cases females achieve more (in acoustic terms) by doing less (in articulatory terms) compared to males.

Acknowledgements

This work was supported by the German Federal Ministry of Education and Research (BMBF) (01UG0711) and two German Research Council Grants (SI 743/6-1,2) awarded to Adrian Simpson. I am grateful to the *Zentrum für Allgemeine Sprachwissenschaft* (ZAS) in Berlin particularly for its financial support regarding the publication of the loop paper, to Jörg Dreyer for technical support with the twin recordings, to the Dept. of Linguistics, University of Potsdam and Adamantios Gafos, Christian Geng and Jana Brunner for help with the gender recordings. Many thanks to the two reviewers of this chapter, the editors of this book, and of course the participating subjects. Thanks also to my marvelous co-authors involved in these studies Leonardo Lancia, Jana Brunner, Susanne Fuchs and Adrian Simpson. All mistakes are my own.

References

Bandura, A. (1977). *Social learning theory*. Englewood Cliffs, N.J.: Prentice-Hall.

Bladon, R., Henton, G., and Pickering, J. (1983). Towards an auditory theory of speaker normalization. *Language and Communication, 4*, 59–69.

Browman, C.P., and Goldstein, L. (1992). Articulatory phonology: an overview. *Phonetica, 49* (3-4), 155–180.

Brunner, J., Fuchs, S., and Perrier, P. (2009). On the relationship between palate shape and articulatory behavior. *The Journal of the Acoustical Society of America*, *125*, 3936–3949.

Brunner, J., Fuchs, S., and Perrier, P. (2011). Supralaryngeal control in Korean velar stops. *Journal of Phonetics* 39(2), 178–195.

Chambers, J. (2003). *Sociolinguistic theory*. Oxford: Blackwell.

Chiba, T., and M. Kajiyama. (1941). *The vowel – its nature and structure.* Tokyo, Japan: Tokyo-Kaiseikan.

Cho, T. (2004). Prosodically conditioned strengthening and vowel-to-vowel coarticulation in English. *Journal of Phonetics 32*(2), 141–176.

Debruyne, F., Decoster, W., Van Gijsel, A., and Vercammen, J. (2002). Speaking fundamental frequency in monozygotic and dizygotic twins. *Journal of Voice* 16(4), 466–471.

de Jong, K. J. (1995). The supraglottal articulation of prominence in English: Linguistic stress as localized hyperarticulation. *The Journal of the Acoustical Society of America, 97*, 491–504.

de Jong, K. J. (1998). Stress-related variation in the articulation of coda alveolar stops: flapping revisited. *Journal of Phonetics, 26*, 283–310.

de Jong, K. J., Beckman, M.E., and Edwards, L. (1993). The interplay between prosodic structure and coarticulation. *Language and Speech* 36 (2-3), 197–212.

Diehl, R. L., Lindblom, B., Hoemeke, K. A., and Fahey, R. P. (1996) On explaining certain male-female differences in the phonetic realization of vowel categories. *Journal of Phonetics, 24*, 187–208.

Eguchi, S., Townsend, G. C., Richards, L. C., Hughes, T., and Kasai, K. (2004). Genetic contribution to dental arch size variation in Australian twins. *Archives of Oral Biology, 49*, 1015–1024.

Fant, G. (1960). *Acoustic theory of speech production*. The Hague: Mouton.

Fant, G. (1966). A note on vocal tract size factors and non-uniform F-pattern scaling, *STL QPSR 4*, 22–30.

Fant, G. (1975). Non-uniform vowel normalization. *STL-QPSR 2-3*, 1–19.

Flanagan, J. R., Ostry, D. J., and Feldman, A. G. (1993). Control of trajectory modifications in target-directed reaching. *Journal of Motor Behaviour, 25*(3), 140–152.

Foulkes, P. and Docherty, G. (2006). The social life of phonetics and phonology. *Journal of Phonetics, 34*, 409–438.

Fuchs, S., Perrier, P., Geng, C., and Mooshammer, C. (2006). What role does the palate play in speech motor control? Insights from tongue kinematics for German alveolar obstruents. In J. Harrington and M. Tabain, (eds.) *Speech production: Models, phonetic processes, and techniques*, New York: Psychology Press, 149–164.

Fuchs, S., Winkler, R., and Perrier, P. (2008). Do speakers' vocal tract geometries shape their articulatory behavior? In *Proceedings of the 8th International Seminar on Speech Production,* Strasbourg, 333–336.

Galton, F. (1876). The history of twins as a criterion of the relative powers of nature and nurture. *Royal Anthropological Institute of Great Britain and Ireland Journal,* 6, 391–406.

Geng, C., Fuchs, S., Mooshammer, C. and Pompino-Marschall, B. (2003). How does vowel context influence loops? In *Proceedings of the 6th International Seminar on Speech Production*, Sydney, Australia, 67–72.

Ghosh, S., Matthies, M., Maas, E., Hanson, A., Tiede, M., Ménard, L., Guenther, F., Lane, H., and Perkell, J. S. (2010). An investigation of the relation between sibilant production and somatosensory and auditory acuity. *The Journal of the Acoustical Society of America, 125*, 3079–3087.

Goldstein, U. (1980). *An articulatory model for the vocal tracts of growing children*. PhD Thesis, MIT.

Gribble, P. L. and Ostry, D. J. (1996). Origins of the power law relation between movement velocity and curvature: Modeling the effects of muscle mechanics and limb dynamics. *Journal of Neurophysiology, 76*(5), 2853–2860.

Gribble, P. L., Ostry, D. J., Sanguineti, V., and Laboissière, R. (1998). Are complex control signals required for human arm movement? *Journal of Neurophysiology, 79*(3), 1409–1424.

Hardcastle, W. J., Gibbon, F., and Nicolaidis, K. (1991). EPG data reduction methods and their implications for studies of lingual coarticulation. *Journal of Phonetics, 19*, 251–266.

Harrington, J. (2006). An acoustic analysis of happy-tensing in the Queen's Christmas broadcasts. *Journal of Phonetics, 34*, 439–457.

Harrington, J., Fletcher, J., and Beckman, M. (2000). Manner and place conflicts in the articulation of accent in Australian English. *In* M. Broe and J. Pierrehumbert (eds.), *Papers in Laboratory Phonology V: Acquisition and the lexicon.* (pp. 40–51) Cambridge: Cambridge University Press.

Henton, C. G. (1995). Cross-language variation in the vowels of female and male speakers. In *Proceedings of the XIIIth International Congress of Phonetic Sciences*, Stockholm, Vol. 4, 420–423.

Hoole, P., Munhall, K. and Mooshammer, C. (1998). Do airstream mechanisms influence tongue movement paths? *Phonetica, 55*, 131–146.

Kabban, M., Fearne, J., Jovanovski, V., and Zou, L. (2001). Tooth size and morphology in twins. *International Journal of Paediatric Dentistry, 11*, 333–339.

Kent, R. and Moll, K. (1972). Cinefluorographic analyses of selected lingual consonants. *Journal of Speech, Language and Hearing Research, 15*, 453–473.

Ladefoged, P. and Broadbent, D. (1957). Information conveyed by vowels. *The Journal of the Acoustical Society of America, 29*(1), 98–104.

Lammert, A.; Proctor, M., and Narayanan, S. (2013). Interspeaker variability in hard palate morphology and vowel production. *Journal of Speech, Language, and Hearing Research 56*, S1924–S1933.

Lancia, L. and Tiede, M. (2012). A survey of methods for the analysis of the temporal evolution of speech articulator trajectories. In S. Fuchs, M. Weirich, D. Pape and P. Perrier (eds.) *Speech production and perception: Speech planning and dynamics*, Frankfurt/Main: Peter Lang, Vol. 1, 239–277.

Langer, P., Tajtáková, M., Bohov, P., and Klimes, I. (1999). Possible role of genetic factors in thyroid growth rate and in the assessment of upper limit of normal thyroid volume in iodine-replete adolescents. *Thyroid, 9*(6), 557–562.

Liberman, A. M., Cooper, F., Shankweiler, D., and Studdart-Kennedy, M. (1967). Perception of the speech code. *Psychological Review, 74*, 431–461.

Liberman, A. M. and Mattingly, I. G. (1985). The motor theory of speech perception revised. *Cognition, 21*, 1–36.

Lindblom, B. (1983). Economy of speech gestures. In P.F. MacNeilage, (ed.) *The production of speech*. New York: Springer, 217–245.

Lindblom, B. (1988). Phonetic invariance and the adaptive nature of speech. *in* B.A. Elsendoom and H. Bouma, (eds.) *Working models of human perception*. London: Academic Press, 139–173.

Lindblom, B. (1990). Explaining phonetic variation: A sketch of the H&H theory. *in* W. J. Hardcastle and A. Marchal, (eds.) *Speech Production and Speech Modelling*, Dordrecht: Kluwer, 403–439.

Loakes, D. (2006). *A forensic phonetic investigation into the speech patterns of identical and non-identical twins.* PhD thesis. University of Melbourne, School of Languages.

Lundström, A. (1948). *Tooth Size and Occlusion in Twins.* Basel: Karger.

Locke, J. L. and Mather, P. L. (1989). Genetic factors in the ontogeny of spoken language: Evidence from monozygotic and dizygotic twins. *Journal of Child Language, 16*(3), 553–559.

Löfqvist, A. and Gracco, V. L. (2002). Control of oral closure in lingual stop consonant production. *The Journal of the Acoustical Society of America, 111*(6), 2811–2827.

Lucero, J. C.; Munhall, K. G.; Gracco, V. L., and Ramsay, J. O. (1997). On the registration of time and the patterning of speech movements. *Journal of Speech, Language and Hearing Research, 40*, 1111–1117.

McGurk, H. and MacDonald, J. (1976). Hearing lips and seeing voices. *Nature, 264*, 746–748.

Mooshammer, C., Hoole, P., and Kühnert, B. (1995). On loops. *Journal of Phonetics, 23*, 3–21.

Mooshammer, C. and Geng, C. (2008). Acoustic and articulatory manifestations of vowel reduction in German. *Journal of the International Phonetic Association, 38*, 117–136.

Nolan, F., Oh, T., McDougal, K., de Jong, G., and Hudson, T. (2006). A forensic phonetic study of 'dynamic' sources of variability in speech: The DyViS project, in *Proceedings of the 11th Australian International Conference on Speech Science and Technology*, Auckland, Australia, 13–18.

Nolan, F. and Oh, T. (1996). Identical twins, different voices. *Forensic Linguistics: International Journal of Speech, Language and the Law, 3*, 39–49.

Nordström, P.-E. (1977). Female and infant vocal tracts simulated from male area functions. *Journal of Phonetics, 5*, 81–92.

O'Neill, B. (2006). *Elementary differential geometry*, 2nd ed., New York: Academic, Chap. 5, 202–263.

Ooki, S. (2005). Genetic and environmental influences on stuttering and tics in Japanese twin children. *Twin Research and Human Genetics, 8*(1), 69–75.

Perkell, J. S. (2010). Movement goals and feedback and feedforward control mechanisms in speech production. *Journal of Neurolinguistics, 25*, 382–407.

Perkell, J. S., Matthies, M. L., Tiede, M., Lane, H., Zandipour, M., Marrone, N., Stockmann, E. and Guenther, F. H. (2004). The distinctness of speakers' /s/-/S/ contrast is related to their auditory discrimination and use of an articulatory saturation effect. *Journal of Speech, Language, and Hearing Research, 47*, 1259–1269.

Perrier, P. and Fuchs, S. (2008). Speed-curvature relations in speech production challenge the 1/3 power law. *Journal of Neurophysiology, 100*(3), 1171–1183.

Perrier, P., Payan, Y., Zandipour, M. and Perkell, J. S. (2003). Influences of tongue biomechanics on speech movements during the production of velar stop consonants: A modeling study. *The Journal of the Acoustical Society of America, 114*, 1582–1599.

Pinheiro, J. and Bates, D. (2000). Mixed-Effects Models in S and S-Plus. Statistics and Computing Series, New York: Springer.

Przybyla, B. D., Horii, J., and Crawford, M. H. (1992). Vocal fundamental frequency in a twin sample: Looking for a genetic effect. *Journal of Voice, 6*(3), 261–266.

R Development Core Team. (2008). *R: A language and environment for statistical computing*. R Foundation for Statistical Computing, Vienna, Austria.

Rudy, K. and Yunusova, Y. (2013). The effect of anatomic factors on tongue position variability during consonants. *Journal of Speech, Language, and Hearing Research, 56*, 137–149.

Ryalls, J. H. and Lieberman, P. (1982). Fundamental frequency and vowel perception. *The Journal of the Acoustical Society of America, 72*, 1631–1634.

Scarr, S. and Carter-Saltzman, L. (1979). Twin method: Defense of a critical assumption. *Behavior Genetics, 9*(6), 527–542.

Simberg, S.; Santtila, P.; Soveri, A.; Varjonen, M.; Sala, E., and Sandnabba, N. K. (2009). Exploring genetic and environmental effects in dysphonia:

A twin study. *Journal of Speech, Language and Hearing Research, 52*, 153–163.

Simpson, A. P. (1998). Phonetische Datenbanken des Deutschen in der empirischen Sprachforschung und der phonologischen Theoriebildung. *Arbeitsberichte des Instituts für Phonetik und digitale Sprachverarbeitung der Universität Kiel* (AIPUK) 33).

Simpson, A. P. (2001). Dynamic consequences of differences in male and female vocal tract dimensions. *The Journal of the Acoustical Society of America, 109*, 2153–2164.

Simpson, A. P. (2002). Gender-specific articulatory-acoustic relations in vowel sequences, *Journal of Phonetics, 30*, 417–435.

Simpson, A. P. and Ericsdotter, C. (2007). Sex-specific differences in f0 and vowel space. In *Proceedings of the XVIth International Congress of Phonetic Sciences*, Saarbrücken, 933–936.

Spinath, F. M. (2005). Twin designs. *In* B. S. Everitt and D. C. Howell, (eds.) *Encyclopedia of statistics in behavioral science*, Chichester: John Wiley & Sons, 2071–2074.

Stevens, K. and Blumstein, S. (1978). Invariant cues for place of articulation in stop consonants. *The Journal of the Acoustical Society of America, 64*, 1358–1368.

Tasko, S. M. and Westbury, J. R. (2004). Speech-curvature relations for speech-related articulatory movement. *Journal of Phonetics, 32*, 65–80.

Toda, M. (2006). Deux stratégies articulatoires pour la réalisation du contraste acoustique des sibilantes /s/ et /ʃ/ en français. *Actes des XXVI ès Journées d'Étude de la Parole*, Dinard, 65–68.

van Lierde, K., Vinck, B., Ley, S., Clement, G., and van Cauwenberge, P. (2005). Genetics of vocal quality characteristics in monozygotic twins: A multiparameter approach. *Journal of Voice, 19*(4), 511–518.

Weirich, M. (2012). *The influence of nature and nurture on speaker-specific parameters in twins' speech: Articulation, acoustics and perception*, Ph.D. dissertation, HU Berlin.

Weirich, M. and Lancia, L. (2011). Perceived auditory similarity and its acoustic correlates in twins and unrelated speakers. In *Proceedings of the XVII International Congress of Phonetic Sciences*, Hong Kong, 2118–2121.

Weirich, M. and Fuchs, S. (2013). Palatal morphology can influence speaker-specific realizations of phonemic contrasts. *Journal of Speech, Language and Hearing Research, 56*, S1894–S1908. [http://jslhr.pubs.asha.org/article.aspx?articleid=1802795]

Weirich, M., Lancia, L., and Brunner, J. (2013). Inter-speaker articulatory variability during vowel-consonant-vowel sequences in twins and unrelated speakers. *The Journal of the Acoustical Society of America, 134* (5), 3766–3780.
[http://scitation.aip.org/content/asa/journal/jasa/134/5/10.1121/1.4822480]

Weirich, M. and Simpson, A.P. (2013). Investigating the relationship between average speaker fundamental frequency and acoustic vowel space size. *The Journal of the Acoustical Society of America, 134* (4), 2965–2974.

Weirich, M. and Simpson, A.P. (2014a). Differences in acoustic vowel space and the perception of speech tempo. *Journal of Phonetics, 43, 1–10.*

Weirich, M. and Simpson, A.P. (2014b). Articulatory vowel spaces of male and female speakers. In *Proceedings of the 10th Intern. Seminar on Speech Production (ISSP)*, Cologne, 453–456.

Whiteside, S. P. (2001). Sex-specific fundamental and formant frequency patterns in a cross-sectional study. *The Journal of the Acoustical Society of America, 110*, 464–478.

Whiteside, S. P. and Rixon, E. (2000). Identification of twins from pure (single) speaker and hybrid (fused) syllables: An acoustic and perceptual case study. *Perceptual and Motor Skills, 91*, 933–947.

Whiteside, S. P. and Rixon, E. (2003). Speech characteristics of monozygotic twins and a same-sex sibling: An acoustic case study of coarticulation patterns in read speech. *Phonetica, 60*, 273–297.

Winkler, R., Fuchs, S., and Perrier, P. (2006). The relation between differences in vocal tract geometry and articulatory control strategies in the production of French vowels: Evidence from MRI and modeling. In *Proceedings of the 7th International Seminar on Speech Production*, Ubatuba, 509–516.

Pascal Perrier[1] and Ralf Winkler[2]

1: Gipsa-lab, CNRS and Grenoble INP
2: Centre for General Linguistics Berlin

Biomechanics of the Orofacial Motor System: Influence of Speaker-Specific Characteristics on Speech Production

Abstract: Orofacial biomechanics has been shown to influence the time signals of speech production and to impose constraints with which the central nervous system has to contend in order to achieve the goals of speech production. After a short explanation of the concept of biomechanics and its link with the variables usually measured in phonetics, two modeling studies are presented, which exemplify the influence of speaker-specific vocal tract morphology and muscle anatomy on speech production. First, speaker-specific 2D biomechanical models of the vocal tract were used that accounted for inter-speaker differences in head morphology. In particular, speakers have different main fiber orientations in the Styloglossus Muscle. Focusing on vowel /i/ it was shown that these differences induce speaker-specific susceptibility to changes in this muscle's activation. Second, the study by Stavness et al. (2013) is summarized. These authors investigated the role of a potential inter-speaker variability of the Orbicularis Oris Muscle implementation with a 3D biomechanical face model. A deeper implementation tends to reduce lip aperture; an increase in peripheralness tends to increase lip protrusion. With these studies, we illustrate the fact that speaker-specific orofacial biomechanics influences the patterns of articulatory and acoustic variability, and the emergence of speech control strategies.

1. Introduction

The variability of speech production observed across native speakers of the same language obviously results from a combination of multiple and complex origins. Among them we can mention social factors such as family origins (Hazen, 2002; Foulkes and Docherty, 2006), gender identity (Fuchs et al., 2010), and sexual orientation (Munson and Babel, 2007), and more intrinsic physical factors such as vocal tract morphology (Fuchs et al., 2008; Winkler et al., 2011a; Lammert et al., 2013) and orofacial biomechanics. In this paper we will focus on the influence of orofacial biomechanics.

With the term *biomechanics*, we understand the *mechanics of the human body*, and with the term *mechanics* we understand:

1) the description of the forces or stresses acting on the body (i.e. the *kinetics* of the body);
2) the characterization of the *intrinsic mechanical* properties of the body, i.e. mass, stiffness, damping, elasticity…;
3) the mathematical formulation of the physical rules determining the link between the forces and stresses applied to the body, and the time motion/deformation of the body; this describes the *dynamics* of the body interacting with its physical environment (see Winters, 2009 for an excellent course about biomechanics and human movements).

Note that the variables characterizing the time motion/deformation of the body, namely its position, velocity and acceleration, are called *kinematic* variables. Kinematic variables are the variables that are usually measured in experimental phonetics. Hence, a biomechanical characterization of speech production goes further into the origins of movements than traditional experimental phonetics.

The quantitative evaluation of the influence of biomechanics on speech articulation in healthy speakers is difficult to achieve. Indeed, when humans or animals produce an intentional movement, their central nervous system (CNS) sends a number of commands to the muscles. These commands, called *motor commands*, will not only generate a displacement of the peripheral motor system (i.e., for example, of the finger, the arm, the limb, the tongue or the mandible), but they will also change some of the biomechanical characteristics of the motor system. It is known for example that the activation of a muscle generates a stiffening of this muscle in the direction orthogonal to the muscle fibers, a phenomenon called *stress stiffening* that is easily observable when someone strongly activates his or her biceps. A sequence of motor commands that achieve a given motor task is called a *motor control strategy*. In speech production healthy speakers have learned how to elaborate motor control strategies of their speech production apparatus in a way that ensures the efficacy of the communication with listeners. Hence, only the result of the combined influences of motor control strategies and biomechanics can be experimentally observed. To evaluate the respective contribution of these two factors individually, it is necessary to design separate models of motor control and biomechanics. These models account for the specific influence of each of these factors on the kinematic properties of the movement. Knowing these separate influences, it is possible to analyze experimental observations from real speakers and reveal how motor control integrates biomechanical constraints to achieve the speech signals that are correctly perceived.

Model-based evaluations of the influence of the dynamics of the vocal tract articulators on speech production have been provided in a number of past studies, in which articulators' dynamics was modeled by a second order system[1]. The authors have in particular investigated the link between articulatory stiffness and clarity of speech production (Browman and Goldstein, 1985; Kelso et al., 1985; Perrier et al., 1996). However, as explained above, biomechanics means much more than dynamics, and the dynamics of the orofacial motor system is much more complex than the one described by a second order system (see Fuchs et al., 2011, for a quantitative evaluation of this specific aspect).

A number of studies have investigated the influence of more complex biomechanical properties of the peripheral motor system on movement trajectories and motor control strategies. Flanagan et al. (1990) have shown that the gently curved shape of the arm trajectories observed in reaching tasks could be the consequence of the motor system dynamics. Perrier et al. (2003) have suggested that the looping patterns observed in tongue movements during the production of [aka], [aku] or [aki] speech sequences (Mooshammer et al., 1995) could arise from a combination of the effects of the dynamics of the tongue and of the muscle arrangements acting on this articulator. Gribble et al. (1996) for arm movements, and Perrier and Fuchs (2008) for tongue movements during speech production, have provided convergent evidence that the relation between trajectory curvature and tangential velocity that is observed in human movements (the so-called 2/3 power law proposed by Viviani and Stucchi, 1992), could result from global dynamical properties of the arm and the tongue. Perrier et al. (2000) have shown that the main directions of tongue deformation for vowels in

1 A *second-order system* is a mechanical system which dynamics is described by a second-order differential equation with coefficients (mass, stiffness, damping) that are constant over time.

various languages (Harshman et al., 1977; Jackson, 1988) correspond to the main directions of the mechanical influences of the synergies between tongue muscles (see also Fuchs and Perrier, 2005).

Nazari et al. (2011) have shown that tissue stiffening in the lips due to the activation of the Orbicularis Oris muscle (see below for more details about this muscle) would significantly help in the achievement of the protrusion and rounding gesture required for the production of /y/ or /u/ in French. Franklin et al. (2007) have experimentally found that in reaching arm movements toward a target the central nervous system (CNS) adjusts muscles' activities so that the arm at target is the least vulnerable to perturbing forces. For this to happen the CNS adjusts the direction of the largest arm stiffness so that it matches the direction along which the reaching task requires the greatest accuracy. In the same vein, Cos et al. (2011) asked human subjects to choose between two potential reaching movements that shared the same ultimate target, but had different characteristics in terms of path distance and mechanical stability at the target. The subjects selected the movements that provided the better stability at the target. Cos et al. (2011) have thus shown that the knowledge of the biomechanical properties of the arm at the target influences decision-making processes in the production of movements.

In North American English, the articulation of the sound /r/ exhibits a noticeable contextual variability for some speakers. In the context of the vowel /i/, /r/ is produced with a bunched tongue having its highest point in the velar region. In the context of the vowel /a/, /r/ is produced with a tip-up tongue shape having its highest point in the alveolar region. Using simulations with a 3D biomechanical model of the tongue (Buchaillard et al., 2009), Stavness et al. (2012) have shown that this co-occurrence can be explained by the fact that it minimizes the change of the stress within the tongue from /r/ to the vowel /i/ or /a/.

Since biomechanics has been shown to influence both motor control strategies and the kinematic properties of movements, it is tempting to think that variability across speakers in the biomechanical characteristics of the orofacial motor system could contribute to the emergence of speaker-specific speech characteristics, also called *speaker idiosyncrasies*. In this paper we will focus on speaker-specific aspects of the *kinetics* of the orofacial motor system. Kinetics includes a description of the mechanisms underlying the generation of muscle forces, and an account of the directions in which these forces are applied. It also integrates the external force field acting on the body. Since most muscles are attached to the skeleton, it is easy to understand that the morphology of the skeleton, namely the size and the shape of the bones, their articulations with each other, i.e. the *anthropometry*, significantly determines the biomechanical properties. This is particularly true for vocal tracts in adults[2]. First, because the shapes of the head and the neck determine the shape of the tongue (Fitch and Giedd, 1999), the direction of the tongue muscle fibers and of their associated forces, and second, because the palate and the tongue interact mechanically through contact forces in particular during consonant production. Hence, in order to study speaker-specific aspects of the kinetics, models have to include a description of the skull and a description of the muscles and muscle force generation mechanisms. Such models are called *biomechanical models*.

In this paper we present two modeling studies based on two kinds of biomechanical models, in which the influence of speaker-specific characteristics will be assessed with simulations. In the first section, some basics in orofacial anatomy will be presented that will facilitate the understanding of the design and the use of the biomechanical models presented in the subsequent sections. In the second section, we will present an assessment of the influence of inter-speaker variations in the global morphology of the skull and neck set on the shape of the tongue and on the control of vowel /i/. This assessment is based on a simplified 2D biomechanical model of the vocal tract, which is adapted to the morphology of two different speakers according to the method proposed by Winkler et al. (2011b). In the last section, an assessment of the influence of potential inter-speaker variations in the Orbicularis Oris anatomy on the lip protrusion gesture will be summarized, which is based on simulations run by Stavness et al. (2013) with a quite complex 3D biomechanical model of the face (Nazari et al., 2010).

2 In children things are more complex since the action of the tongue on the palate during swallowing and perhaps speech production largely influences the final palatal shape, and because tongue movements in general contribute to the evolution of the vocal tract during vocal tract growth.

2. Some basics in orofacial muscle anatomy

In this section we provide basic information about the anatomy of the tongue and face that will be useful for understanding the modeling work presented below. This description contains a number of simplifications of the quite complex anatomical reality. For the tongue a very accurate description can be found in Takemoto (2001).

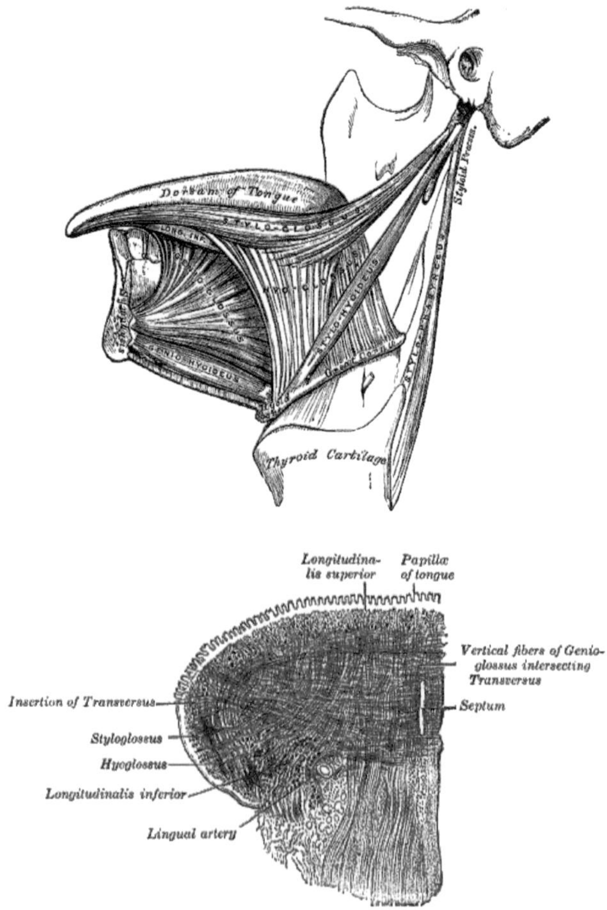

Figure 1: Representation of the main muscles acting on the mobile part of the tongue. Upper panel: view from the left hand side; bottom panel: transversal cut of half the tongue (from the left to the right) seen from the front (from Gray, 1918 in Bartleby.com, 2000).

The mobile part of the tongue is controlled by eight muscles that are represented in Figure 1. Four of these muscles are considered to be *extrinsic* muscles, because at one of their extremities they attach to structures that are external to the tongue. They are as follows: *Genioglossus*, in the central part of the tongue, which originates from the inner mandibular surface at the Symphysis (bottom left of the top panel); the *Styloglossus* which emanates from the styloid process in the temporal region of the head (upper right of the top panel); the *Hyoglossus* originating from the greater horns of the hyoid bone (bottom right of the top panel); and the *Palatoglossus* (not represented in this figure) emanating from the anterolateral palatal aponeurosis in the soft palate. The other four muscles are *intrinsic*, since both extremities are within the tongue (see in particular the bottom panel in Figure 1): the *Longitudinalis Superior*, the *Longitudinalis Inferior*, the *Transversus* and the *Verticalis* (not represented in this figure). Not listed here, other muscles located in the mouth floor act indirectly on the tongue, in particular muscles involved in hyoid bone movement. The fiber directions of the extrinsic muscles are influenced by the shape of the tongue and also by the morphology of the jaw, the hyoid bone and the temporal bone, while fiber directions of the intrinsic muscles only depend on the tongue shape.

The lip shape can be modified by the control of 11 orofacial pairs of muscles (see Figure 2) located symmetrically on both sides of the midsagittal plane. According to their influence on the lips, they are classified into the upper lip elevators (*Levator Labii Alaeque Nasi*, *Levator Labii Superioris* and *Zygomaticus Minor*), the lip corner mobilizers (*Levator Anguli Oris*, *Zygomaticus Major*, *Risorius*, *Buccinator* and *Depressor Anguli Oris*), the lower lip mobilizers (*Depressor Labii Inferioris and Mentalis,* not represented in Figure 2) and the oral fissure constrictors (*Peripheralis* and *Marginalis parts of the Orbicularis Oris*). All these muscles originate from bony structures of the skull, except the Orbicularis Oris muscle, which emanates from a lip corner and inserts into the opposite corner of the lips (muscular tissue). The Orbicularis Oris muscle is composed of an upper and a lower part.

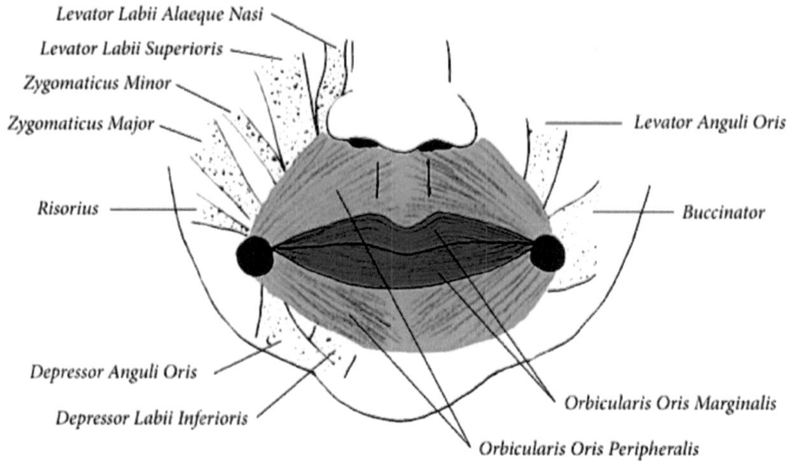

Figure 2: Schematic representation of the muscles determining the shape of the lips. These muscles are grouped in pairs located symmetrically on both sides of the mid-sagittal plane, but for matter of simplification only one side of each muscle pair is represented. Reprinted from Journal of Anatomy 214*(1), 36-44, Rogers, C.R., Mooney, M. P., Smith, T. D., Weinberg, S. M., Waller, B. M., Parr, L. A., Docherty, B. A., Bonar, C. J., Reinholt, L. E., Deleyiannis, F. W.-B., Siegel, M. I., Marazita, M. L., and Burrows, A. M. Comparative microanatomy of the orbicularis oris muscle between chimpanzees and humans: evolutionary divergence of lip function. Reproduced with permission from John Wiley and Sons. Copyright 2008.*

3. Inter-speaker variation in extrinsic tongue muscles orientation

In this section a 2D biomechanical model of the tongue is used to assess the impact of inter-speaker variations in head and neck morphology on the tongue muscle fibers' directions and on the patterns of articulatory and acoustic variability in the production of the high front vowel /i/. Vowel /i/ has been chosen for this evaluation because its production requires precise tongue positioning.

3.1. Methodology

We used the 2D biomechanical model of the tongue developed by Payan and Perrier (1997) in its most recent version (Perrier et al., 2003). It mainly consists of a deformable Finite Element Mesh (FEM) embedded in rigid vocal tract walls in the mid-sagittal plane. The 2D mesh is a simplified representation of the 3D tongue structure. It is considered to be a projection of the tongue in the mid-sagittal plane. The geometry of the mesh (see Figure 3) is specifically designed to facilitate the anatomical representation of the muscles acting on the position and shape of the tongue in the front-back direction. The external contour of the mesh was derived from an X-ray view of the vocal tract of a male speaker at rest (close to a schwa production). Five muscles are represented: the Genioglossus, the Styloglossus, the Hyoglossus, the Verticalis and the Longitudinalis Inferior. The Genioglossus has been divided in two functional parts, the Posterior and the Anterior Genioglossus. Muscle activations are controlled according to the λ-model (Feldman, 1986), which generates a force for each muscle that is a function of the difference between the motor control variable λ specified for this muscle and the actual muscle length. If the actual length is smaller than λ no active muscle force is generated. If the actual muscle length is equal to or larger than λ the force develops as an increasing function of the actual muscle length. In sum, in a given static position of the tongue, in which a muscle M has the length l, the force F_M generated by the muscle varies with the motor control variable λ according to the equation:

$$F_M = \rho.(e^{c(l-\lambda)} - 1), if\ l \geq \lambda, and\ F_M = 0\ if\ l \leq \lambda \qquad (1)$$

where c is a form parameter and ρ is an amplitude parameter directly related to the force generation capacity of the muscle (for more details see: Laboissière et al., 1996; Payan and Perrier, 1997). λ can be seen as the threshold muscle length above which muscle force starts developing. In spite of its simplicity this 2D biomechanical model has been shown to be capable of accounting for some important kinematic characteristics of speech articulation, which have been experimentally observed in different speakers of different languages: velocity profiles (Payan and Perrier, 1997), trajectory shapes (Perrier et al., 2003), or relations between trajectory curvature and speed (Perrier and Fuchs, 2008).

This model serves as a reference model, from which speaker-specific 2D biomechanical models can be routinely developed according to the method proposed by Winkler et al. (2011b). Two basic hypotheses underlie the adaptation of the model to a specific speaker: (1) the general anatomical arrangements accounted for by the mesh geometry in the reference model is common to all human beings, (2) variations across speakers in muscle lengths and muscle orientations within the tongue are strongly correlated with global variations of the head and neck morphology, such as variations in larynx height, length of the mandible ramus, head size, and mid-sagittal palate shape. Taking these assumptions into account, the transformation of the reference model requires contours reflecting the vocal tract morphology and anatomical landmarks corresponding to muscle fiber origins. The two contours are the tongue contour at rest, and the mid-sagittal external contour including the upper lip, the palate, the soft palate and the pharyngeal walls. The three landmarks are the lower (P1) and upper (P2) limit of the tongue where the Genioglossus emanates from the mandibular Symphysis, and the Styloid process (P3) (see Figure 3 for a representation of these landmarks on the reference model).

Once these anatomical landmarks are determined on the speaker (see below for details), the generation of the speaker-specific biomechanical model is straightforward. First, the upper contour of the tongue model is projected onto the mid-sagittal tongue contour measured for the subject. Second, the distribution of the nodes along this new upper contour is made proportionally to the distribution of the nodes in the reference model. Third, the lower and upper attachment points of the new tongue mesh on the mandible are assigned to points P1 and P2. Then, the distribution of the nodes within the mesh is obtained by deforming the original mesh linearly from the nodes on the upper contour to the insertion nodes P1 and P2 of the mesh into the mandibular Symphysis. The difference in size and orientation of the segment [P1 P2] between the reference model and the speaker-specific morphology serves to transform the size and the orientation of the incisor representing the mandible in the sagittal plane. Finally, the extremity of the external Styloglossus fiber is attached to point P3. This matching procedure fully determines the geometry of the new mesh and consequently the muscle arrangement within the speaker-specific tongue model. It preserves the original topology of the mesh while accounting for the speaker-specific morphology.

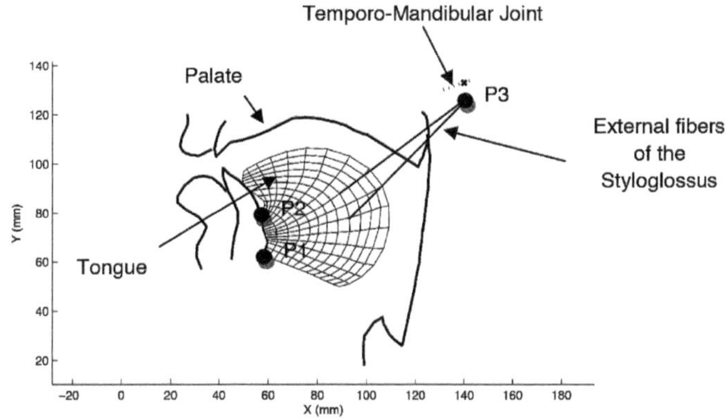

Figure 3: The reference 2D biomechanical model of the vocal tract. The tongue and jaw position correspond to the positions observed at rest for the reference subject with X-Ray imaging from the side. The anatomical landmarks that serve as a basis for the transformation of this model into a speaker-specific model are indicated.

Just as in the reference model, the speaker-specific tongue mesh obtained after the adaptation procedure represents the tongue at rest. The lengths of the muscles in this rest configuration determine speaker-specific reference λ commands. For all speakers, if the λ commands are equal to the reference values, the force generated by each muscle in the rest configuration is equal to zero. These reference λ commands are used to establish the correspondence between the motor commands used in the speaker-specific models and in the reference model: we consider the motor commands to be equal in all the models if the difference δλ between the actual λ values and the speaker-specific reference λ commands are equal.

In order to generate the acoustic signal associated with a given vocal-tract configuration, the 2D mid-sagittal representation of the vocal tract has to be converted to its corresponding area function. This is accomplished first by determining the variation of the mid-sagittal distance from the glottis to the lips. Then, the area function is computed from the sagittal distance by applying an enhanced version of the model proposed by Perrier et al. (1992). The mid-sagittal distance is measured on a grid that is projected on the geometry of the biomechanical model. For the speaker-specific model a grid derived from the grid proposed in Perrier et al. (1992) is used. The grid is divided into a pharyngeal section from the glottis to the velum, and a palatal section from the velum to the lips. The interval between the lines of the grid and the angle between the pharyngeal and the palatal part have been adapted in order to match the length and angle characteristics of the speaker to whom the model is adapted. Then, the exact same procedure was applied for all the models to compute the area function from the sagittal distance. Doing so, we do not account for inter-speaker differences in the transversal direction, i. e. the direction orthogonal to the mid-sagittal plane. This choice is justified by the fact that we want to only assess inter-speaker differences associated with the biomechanical specificities accounted for in the model.

Finally, the acoustic signal is generated from the area with a reflection-type line analog of the vocal tract (Story et al., 2000). Vocal folds oscillations are generated and controlled with a numerical implementation of the three-mass model designed by Story and Titze (1995) based on lumped-elements (Titze and Story, 2002).

In order to illustrate with this procedure the potential impact of speaker-specific biomechanics on speech production, we have focused on the production of vowel /i/. Vowel /i/ is interesting for three main reasons: (1) it is an extreme vowel that exists in all the languages of the world (Ladefoged and Maddieson, 1996); (2) the correct acoustic realization of this vowel requires a precise position of the tongue along the palate (Gay et al., 1992); and (3) the articulation of this vowel requires mainly the activation of the posterior Genioglossus and the Styloglossus (see for example Buchaillard et al., 2009), two muscles that are likely to be significantly impacted by the variation of the head and neck morphology across speakers. We have focused on the variation in articulation and in acoustics associated with local variations of the activations of the Posterior Genioglossus, the Styloglossus, the Anterior Genioglossus and the Hyoglossus. These muscles have been shown to be the most important for tongue position control in vowel production (Honda, 1996).

We first determined for each model a tongue configuration corresponding to a prototypical /i/. This prototype was obtained in two steps. First, 1000 tongue configurations were generated by a random sampling of the 6-dimensional space of the motor control variables (the λ-space), expressed in terms of their differences with the reference λ commands in the tongue rest position. Among these 1000 configurations one configuration was selected for which the formant patterns and the sagittal view of the model corresponded to the vowel /i/. Second, starting from this /i/ configuration, we adjusted step-by-step the λ values of the Posterior Genioglossus muscle and the Styloglossus muscle in order to improve the characteristics of the vowel /i/. The criteria are that a prototypical /i/ is characterized by a narrow constriction in the alveolar region and by the highest possible value of the second formant. For each model our standard /i/ configuration had these two basic characteristics. For each model different articulatory configurations were generated around the corresponding prototypical /i/ configuration, by changing the motor control variables to the Posterior Genioglossus, the Styloglossus, the Anterior Genioglossus and the Hyoglossus within a range of variation of δλ, the difference between the actual λ values and the reference λ values commands at rest, equal to [-2 mm +2 mm] with a 1-mm-step. Thus, five different λ values have been used for each muscle and all the combinations of the λ values of the four muscles were used ($5^4 = 625$ articulatory configurations). Finally the variation in the sagittal plane and in the acoustic domain was assessed and compared across speakers.

This methodology was applied to two speakers, a female speaker S1 and a male speaker S2. These two speakers were selected from a set of 13 subjects for whom MRI anatomical data were available (Apostol, 2001), because they are quite representative of the vocal tract differences between female and male. The results of the simulations obtained for these two models and for the reference model are presented and analyzed.

3.2. Results

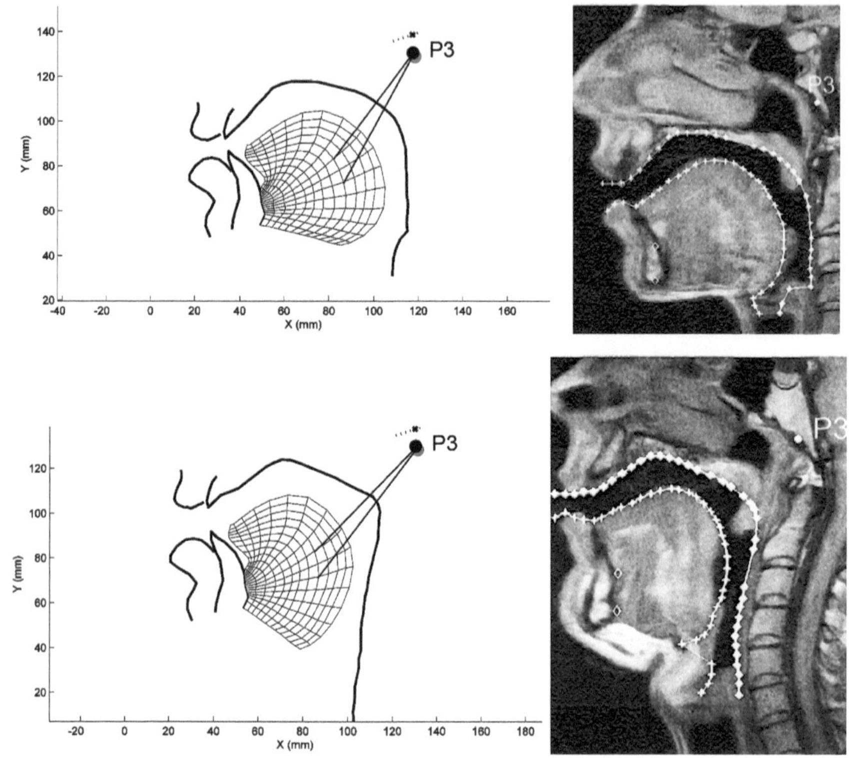

Figure 4: Adapted 2D biomechanical models of the vocal tract (left panels, lips are on the left hand side) and mid-sagittal MR images of the vocal tract at tongue rest position (right panels), for subjects S1 (top) and S2 (bottom). For comparison with the Reference Model see Figure 3.

The geometrical transformation of the reference model into the speaker-specific model induces changes in the direction of the muscle fibers. In Figure 5, we can observe these changes for the two main muscles involved in the production of vowel /i/, the Posterior Genioglossus (left) and the Styloglossus (right). For the Posterior Genioglossus few differences are observed between S1 and the reference model; for S2 the lower fibers of this muscle are more inclined than in the reference model. For the Styloglossus muscle the two speakers S1 and S2 present external fibers that are clearly more vertical than in the reference model. This phenomenon is stronger for S1 than for S2. Accordingly we expect the Styloglossus muscle to generate movements of the tongue that are more vertical and less horizontal in S2 than in S1 as well as than in the reference model.

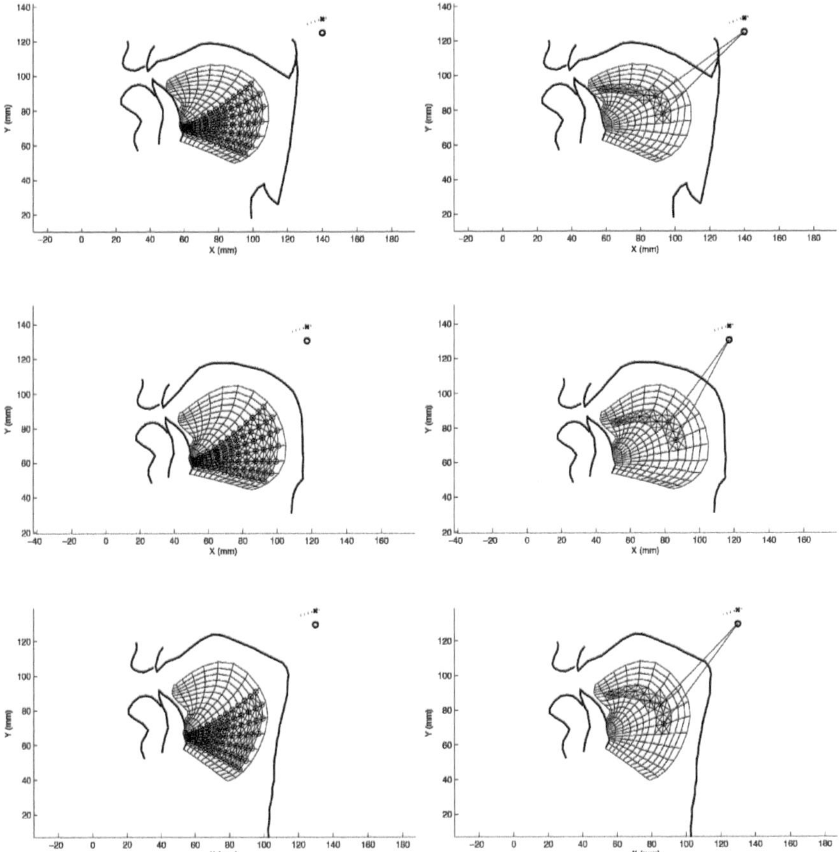

Figure 5: Fibers' implementation of the Posterior Genioglossus (left panels) and of the Styloglossus (right panels) in the 2D biomechanical models of the vocal tract. The circles on the edges of the elements of the mesh describe the path of the fibers in the mesh. The solid lines joining the Styloid process (circle on the upper right corner of each panel) represent the fibers that are external to the tongue. The crossed elements in the mesh correspond to the muscle body. Their stiffness increases when the muscle is activated. From the top to the bottom: reference model; Subject S1; Subject S2.

In Figure 6, the vocal tract configuration selected for vowel /i/ is represented for each model. All of them have a constriction in the alveolar region, but the length of the constriction along the front/back direction varies across the models, due to differences in tongue shapes and in palatal contours. Speaker S1 seems to have a longer constriction than speaker S2, and the reference model. This is confirmed by the computation of the area functions (see Figure 7). The force produced by each muscle in this configuration was computed as described in equation 1. For the reference model the ratio between the force exerted by the Posterior Genioglossus and the one exerted by the Styloglossus is equal to 0.75. The same ratio is found for S1, but this ratio is equal to 1.9 for S2. This difference is consistent with the fact that at rest the tongue and jaw are lower, i. e. the tongue is further apart from the occlusal plane for S2 than for S1 and the reference model (see Figures 3 and 4).

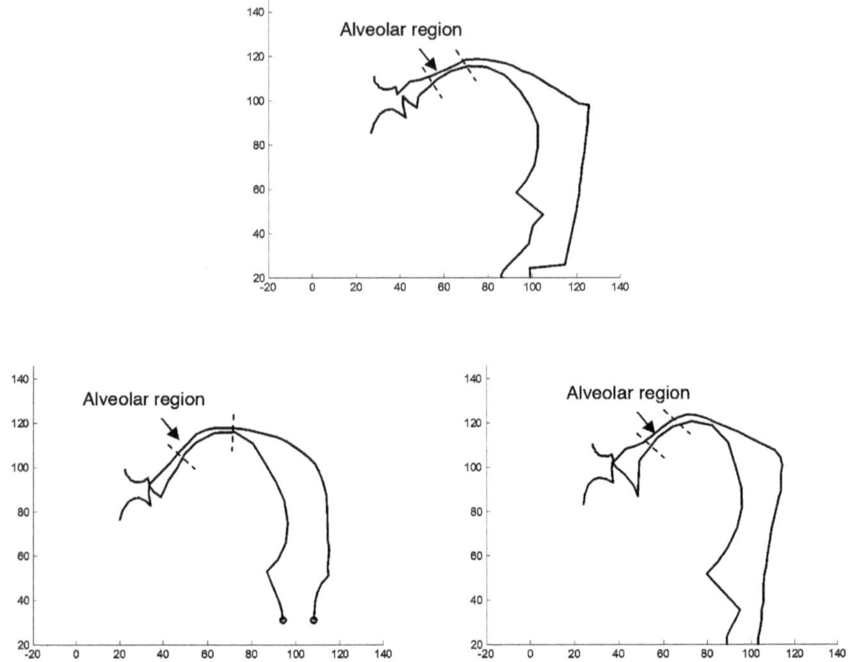

Figure 6: Mid-sagittal views of vowel /i/ generated by the speaker-specific 2D biomechanical models and the Reference Model of the tongue (see Figure 5). The dotted lines show rough estimations of the constriction's boundaries. Top panel: Reference Model. Bottom panels: Left, subject S1; Right: subject S2. Lips are on the left hand side.

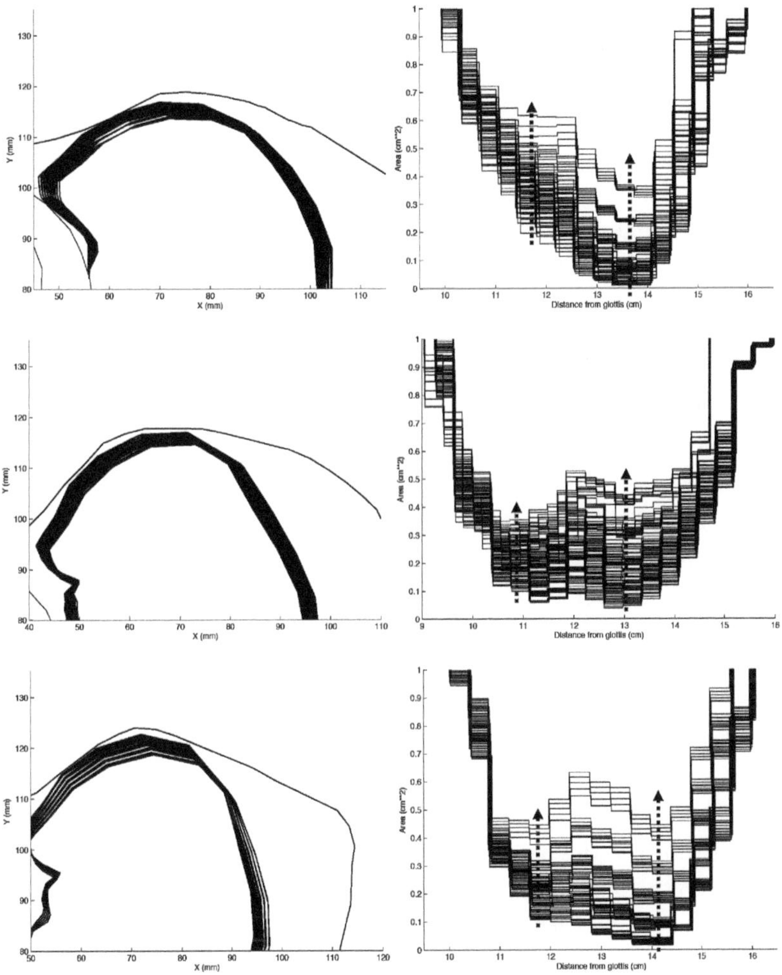

Figure 7: Variations of the articulatory configuration observed for vowel /i/ when the activation of the Styloglossus varies. The left panels show tongue positions in the mid-sagittal plane, in the region of the constriction (lips are on the left hand side). The right panels show the variation of the area function in the region of the constriction (front is on the right hand side). The dotted arrows superimposed on the plots of the area functions give the main directions of the area changes in the constriction's region. The size of the arrows corresponds to the amplitude of the area change. From the top to the bottom: Reference Model; Subject S1; Subject S2.

Figure 7 presents the results of the random variation of the motor control variables λ to the four main muscles, according to the methodology described above. The left panels represent the tongue contour variations in the mid-sagittal plane with a focus on the palatal region. The right panels represent the corresponding variations of the area function, focusing on the region of the constriction. The main direction of the tongue contour variations changes across the models. For the reference model, the variation in the mid-sagittal plane is essentially along a front/high-back/low direction. This is associated with a change in the constriction opening. For speaker S1, the variation in the mid-sagittal plane is two-fold: the variation of the opening of the constriction in the alveolar region is significantly larger than the variation of the constriction in the post-alveolar region. The consequence for the area function is that the opening/closing of the front part of the constriction is associated with a relative closing/opening of the back part of the constriction. Thus, variations in the main muscle activations are associated with a change in the main constriction location. For speaker S2, the pattern of variation is intermediary between speaker S1 and the reference model. The main trend is a global opening of the constriction, but the narrowest part of the constriction moves backwards. A detailed analysis of the effects of the four different muscles taken separately has revealed that these patterns of variation are mainly associated with the action of the Styloglossus muscle. This statement is consistent with the observations of the differences existing across speakers in the direction of the external fibers of the Styloglossus, as observed on Figure 5: for speaker S1, the orientation of these fibers is more vertical than for S2 and the reference model, and the vertical component associated with changes in muscle activations is stronger. The increase in Styloglossus activation creates for speaker S1 a constriction just behind the original place of constriction. A similar trend exists also in S2 as compared to the reference model but it is smaller.

The acoustic variations associated with the articulatory variations shown in Figure 7 are depicted in Figure 8. Note that the scaling of the figures is the same for the three models. The differences in the main orientations of the dispersion ellipses across models inform about the main impact in the acoustical domain of the biomechanical differences.

For S1 the variability along the F3 dimension is clearly stronger and the variability along the F1 direction clearly smaller than for S2 and, to a lesser extent, the reference model. This is consistent with the fact that in S1 the constriction location moves along the front/back direction due to the orientation of the force exerted by the Styloglossus relatively to the palatal contour, while its cross-section changes less than for the other two models. S2 has the largest variability in the (F2, F1) space and the smallest variability along the F3 dimension. The reference model is intermediary. For a correct perception of vowel /i/, reaching a high F3 value is important (see for example Schwartz et al., 1993). Hence, these simulations suggest that model S1 requires a more accurate control of the Styloglossus muscle activation than the remaining two models.

Obviously, for a comprehensive analysis, the influences of the other muscles should be taken in consideration. We can also not discard the possibility that our observations are linked with the special standard configuration chosen for vowel /i/, even if the results are very consistent with the differences observed in the Styloglossus fibers' orientation across speakers. It is not possible to draw strong conclusions from this limited study. Our results just aim to illustrate how speaker-specific biomechanics can influence motor control strategies and could explain in part some trends in idiosyncrasies.

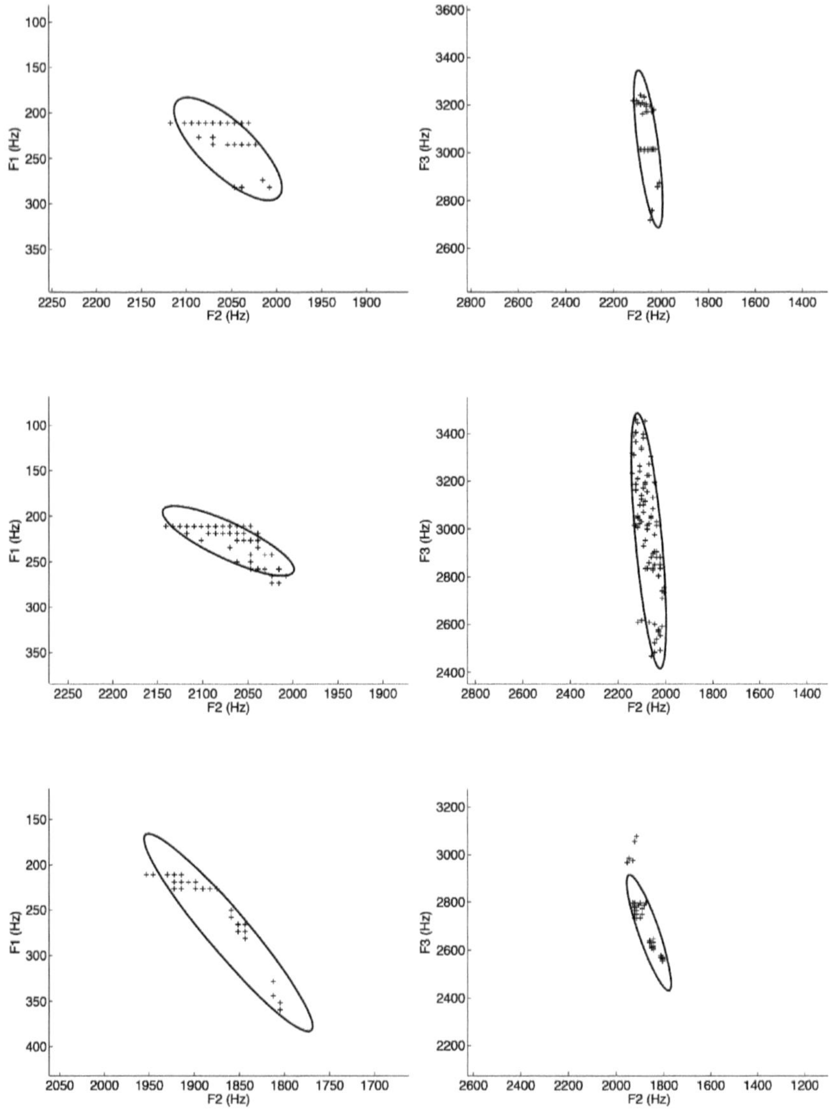

Figure 8: Variability in the (F2, F1) (left panels) and (F2, F3) planes (right panels) associated with local variations of the four main muscle activations for vowel /i/. The ellipses represent the 2 σ dispersion ellipses inferred from the data dispersion assuming a normal distribution. From the top to the bottom: Reference model; Subject S1; Subject S2.

4. A modeling study of anatomical variability in Orbicularis Oris

In articulatory phonetics lip protrusion is considered to be the basic gesture underlying the production of rounded vowels such as /u/ or /o/. The acoustic characteristics of rounded vowels as compared to unrounded or spread vowels are well-described and consistent in many languages. They correspond to an increase of the spectral energy in the low frequencies and a decrease in the high frequencies. However, the actual gesture underlying the production of rounded vowels can significantly vary across speakers. For a large part of the speakers the lips are protruded to the front and the lip orifice has a small area and is round. For another part of the speakers the lips are not protruded; the lip orifice has a small area but it is not round. Stavness et al. (2013) provided two characteristic examples of these two different articulatory strategies (see their Figures 1 and 2, p. 879). Stavness et al. (2013) have investigated the potential contribution of anatomical variability in the distribution of the muscle fibers between the Peripheralis and the Marginalis parts of the Orbicularis Oris. Facial muscles present a non-negligible variability across humans. Stavness et al. (2013) cited for example the studies of Huber (1933) who found that the Risorius muscle (see Figure 1) exists in only 20% of the Melanesians and in 80% of the Europeans. They also referred to Pessa et al. (1998) who observed that among the 50 specimens that they studied, 17 presented a Zygomaticus Major muscle with a bifid structure, i.e. with two insertion points on the skull. This peculiarity could be responsible for the dimple in the cheeks that many people have when smiling. To our knowledge no study has shown that significant differences exist among humans in the morphology of the Orbicularis Oris muscle. Nevertheless, since the emergence of distinct Marginalis and Peripheralis parts in this muscle seem to be quite recent in the primates' development (Rogers et al., 2009), it is not unlikely that a variability exists. Citing Ladefoged (1984), Stavness et al. (2013) suggest that such variability would be consistent with the fact that individual differences in facial mimics are compatible with individual differences in lip shaping during speech production.

The investigation was based on simulations run with a sophisticated 3D Finite Element biomechanical model of the face (Nazari et al., 2010, 2011;

Stavness et al., 2014). This model includes a 3D anatomical representation of all the muscles that are mentioned in section 1 and displayed in Figure 2. Muscle mechanics is accounted for with a Finite Element model of the Hill-type muscle model (Blemker et al., 2005). Details about the parameterization of the muscle model can be found in Stavness et al. (2013). The Orbicularis Oris muscle is represented as a continuous loop of elements around the labial orifice as depicted in Figure 9.

In order to evaluate the influence of the anatomical variability in this muscle, Stavness and colleagues performed two different sets of simulations:

1. To evaluate the influence of the depth of the muscle implementation they considered simulations with active elements located only in the deep (D), or in the middle (M), or in the superficial (S) layer of the mesh (see Figure 9, bottom right panel)
2. To evaluate the influence of the size of the muscle implementation they considered simulations with active elements of various sizes, from marginal to peripheral (1, 2, 3, 4 on Figure 9, bottom right panel).

Simulations were performed in ArtiSynth (http://artisynth.magic.ubc.ca/artisynth/), which is a 3D platform for fast-forward dynamics simulation with dynamic coupling between rigid body and soft Finite Element models as well as collision handling. Each simulation was 500 ms in duration. Muscle activation increased linearly up to 400 ms and held the final activation for 100 ms. In all simulations, muscle activation was increased uniformly from 0% to 50% of the maximum possible activation, which corresponds to an active muscle stress of 50 kPa. This level of final activation was chosen to ensure numerical convergence in all simulations while generating lip displacements of realistic amplitudes. Each simulation reached an equilibrium position by 500 ms.

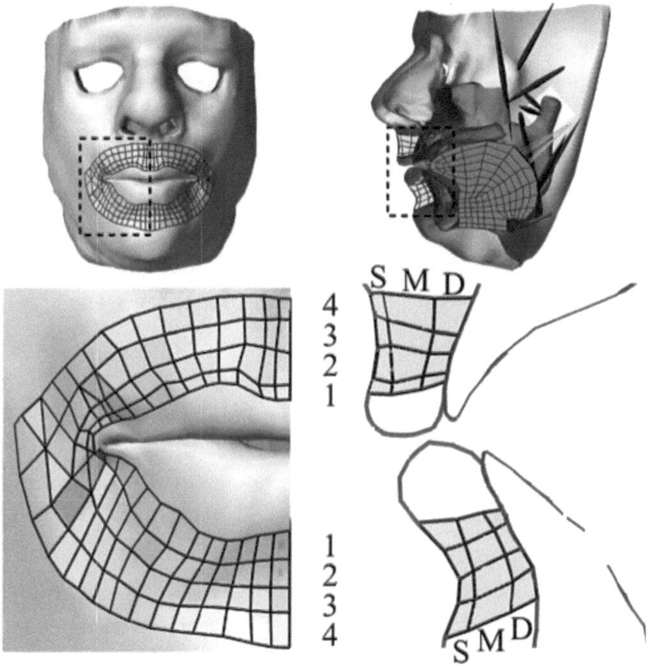

Figure 9: Front (left panels) and side (right panels) views of the face model showing the Orbicularis Oris muscle elements organized into different peripheral loops from marginal to peripheral (1, 2, 3, 4), and into different depth layers in the mesh, superficial (S), middle (M), and deep (D). Reprinted from Journal of Speech, Language, and Hearing Research, *56(3), 878–890, Stavness, I., Nazari, M. A., Perrier, P., Demolin, D., and Payan, Y. A biomechanical modeling study of the effects of the orbicularis oris muscle and jaw posture on lip shape. Reproduced with permission from the American Speech-Language-Hearing Association (http://jslhr.pubs.asha.org). Copyright 2013.*

The results of the different simulations are summarized in Figure 10. Each row represents how lips shape varies, for a given deepness, when the implementation changes from marginal to peripheral. Each column shows the influence of deepness, from superficial to deep, for a given peripheralness. In each panel a front view of the lip horn is presented on the left and a side view is presented on the right. Protrusion can be seen on the side view. The area and the shape of the lip orifice can be seen on the front view. We can see that large lip shape variability is associated with the implemented anatomical variability. A deep implementation tends to reduce the vertical dimension of the labial orifice. This is probably due to the fact that the contraction of the deep part of the muscle generates an inward displacement of the whole labial tissue, while a superficial implementation only acts on the superficial labial tissues. More peripheral implementations are associated with larger lip protrusion. This can be explained by the combination of the effects of the teeth, as a rigid obstacle, and the quasi-incompressibility of the labial tissues. Labial tissues tend to maintain their global volume quasi constant whatever the stress applied to them. When the labial tissues are compressed in the region close to the teeth (the more peripheral one), the compensatory expansion of the volume in the other parts of the tissues can only occur in the front direction, since the teeth block the expansion in the back direction. For the other degree of peripheralness, the volume expansion can occur in both directions. Interestingly, deepness influences the impact of the degree of peripheralness on lip aperture: for a superficial implementation, lip aperture varies monotonously with the increase of peripheralness; for a middle implementation the aperture increases from peripheralness degree 1 to degree 3, and decreases from degree 3 to degree 4; for a deep implementation the peripheralness has little impact on the very small lip aperture. The most prototypical lip shape corresponding to a French rounded vowel like /u/ or /y/ is only observed for a middle deepness and a middle peripheral (levels 3 and 4) implementation of the Orbicularis Oris.

This set of simulations illustrates how individual variation in Orbicularis Oris muscle anatomy could influence the gesture underlying the production of rounded vowels. In subjects having a quite marginal Orbicularis Oris implementation, it seems more difficult to generate a protrusion of the lips and to achieve a small round lip orifice. Again biomechanics determines the constraints applied to the achievement of a gesture, and the central nervous system can elaborate different motor control strategies to deal with these constraints. Hence, it is possible that a lip protrusion and a round lip orifice are achieved in spite of a marginal implementation of the Orbicularis Oris. However, such marginal implementation could make these gestures more complex, with the consequence that they would be observed less often than in subjects with a more peripheral implementation.

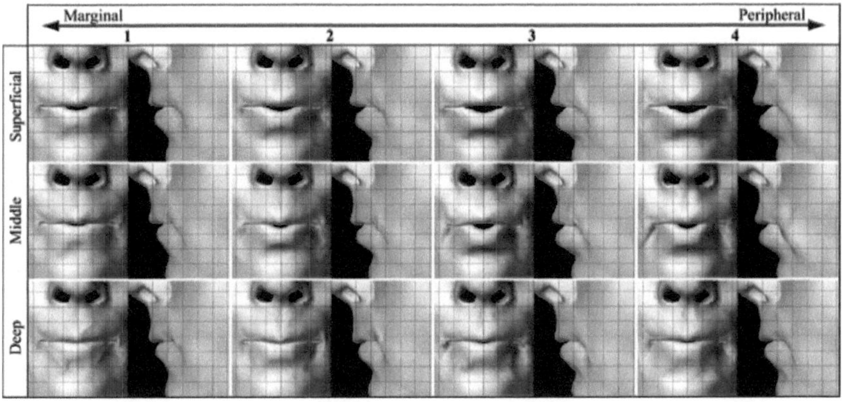

Figure 10: Simulation results for different Orbicularis Oris muscle deepness and peripheralness. Reprinted from Journal of Speech, Language, and Hearing Research, *56(3), 878–890, Stavness, I., Nazari, M. A., Perrier, P., Demolin, D., Payan, Y. A biomechanical modeling study of the effects of the orbicularis oris muscle and jaw posture on lip shape. Reproduced with permission from the American Speech-Language-Hearing Association (http://jslhr.pubs.asha.org). Copyright 2013.*

5. Conclusions

The rare studies of the influence of individual biomechanical factors on subject-specific motor control strategies in very skilled motor tasks have shown that this influence is limited. Frère and Hug (2012) have studied nine high level gymnasts with different morphologies during backward giant swings in the high bar. They have computed the correlations between their different muscle activities in order to extract synergies, independently for each gymnast. They found that the nine subjects share the same first two main synergies. Differences started to be significant only from the third most important synergy. A similar observation was done by Hug et al. (2010), who studied muscles synergies in eleven highly trained cyclists in an experimental protocol in which the torque they had to counteract, the torque-velocity relation, and their posture varied significantly. These authors found that the three first synergies remain the same across conditions. Since speech production is also a highly skilled motor task we expect similar findings. We believe that the most known synergies observed in speech production are certainly shared by the huge majority of humans.

We believe that biomechanical influences are more subtle and affect the balance within the synergies, and the sensitivity of the articulatory configuration to small variations in muscle activation. Thus, speaker-specific biomechanical properties can influence the level of accuracy required for the production of given sounds.

With simulations performed with two kinds of biomechanical models of the orofacial motor system, we have shown examples of the potential influences of speaker-specific biomechanics on the production of speech gestures. These examples show how inter-speaker differences in muscle anatomy can generate inter-speaker differences in motor control strategies or/and in articulatory and acoustic variability. Work is currently in progress in our lab to assess how these phenomena could influence coarticulation strategies. Coarticulation strategies determine the way gestures are organized, sequentially and in parallel, for the production of a speech sequence. Coarticulation strategies use the degrees of freedom of the speech production system to optimize the gestures while preserving the ultimate goal of speech production – its correct perception by listeners (Whalen, 1990; Lindblom, 1990). The example of the speaker-dependent impact of variations in the muscle activations around vowel /i/ (section 3) on articulatory and acoustic variability has shown how biomechanics can change the degrees of freedom and the accuracy in the achievement of a given speech task. The study of the impact of the Orbicularis Oris implementation on the production of rounded vowels (section 4) suggests that biomechanics can change the motor control strategies underlying the production of speech. With these two limited examples we do not pretend to cover all the ranges of the potential influences of biomechanics on speech motor control. We have shown that orofacial biomechanics can influence the emergence of motor strategies in speech production, due to the fact that it affects the degrees of freedom and the accuracy of the control. Coarticulatory variability results to a large extent from the use of the degrees of freedom to anticipate forthcoming gestures and reduce speech effort, while preserving a satisfactory accuracy to enable a good perception of the speech signal. Hence, it is likely that idiosyncrasies originate in part in speaker-specific biomechanical factors.

This could have an influence not only on speech production, but also on speech perception. It has been shown that an interaction exists between the motor control underlying the production of the sounds and perceptual boundaries for these sounds. For example Shiller et al. (2009) have perturbed the auditory feedback of speakers during the production of the fricative /s/, in order to make it sound more like a /ʃ/. To do so they shifted the spectral energy toward the low frequencies. They observed that the subjects tend to correct their articulation in order for the corrected articulation to generate a perceived sound that is closer to their usual /s/, in spite of the perturbation of the auditory feedback. The subjects produced a more anterior articulation of /s/, in order for the spectral energy to move back to the high frequencies. Interestingly a perceptual test run immediately after this experiment has shown that the perceptual boundary between /s/ and /ʃ/ has moved: the subjects tolerate more low frequencies for /s/ than before the experiment. This result suggests that in presence of the perturbed auditory feedback, due to the influence of the usual articulation of /s/ and /ʃ/, the subjects have limited the articulatory changes and accepted a small shift in their perceptual boundaries. Since motor control seems to influence perceptual classes, we expect that the articulatory variability compatible with a correct perception of a sound could influence the tolerated perceptual variability. Thus, we can imagine a scenario in which idiosyncrasies would emerge from the interaction between biomechanical constraints, perceptual accuracy and social and cultural influences.

Acknowledgements

This work was supported by the German Research Council to the SPEECHart project (Grant Nr. FU 791/1-1)

References

Apostol, L. (2001). *Étude et simulation des caractéristiques individuelles des locuteurs par modélisation du processus de production de la parole.* Unpublished Doctoral dissertation, Grenoble: Institut National Polytechnique de Grenoble.

Blemker, S. S., Pinsky, P. M., and Delp, S. L. (2005). A 3D model of muscle reveals the causes of nonuniform strains in the biceps brachii. *Journal of Biomechanics*, *38*(4), 657–665.

Browman, C.P., and Goldstein, L. (1985). Dynamic modeling of phonetic structure. In V. Fromkin (ed.). *Phonetic Linguistics* (pp. 35–53). New York: Academic Press.

Buchaillard, S., Perrier, P., and Payan, Y. (2009). A biomechanical model of cardinal vowel production: Muscle activations and the impact of gravity on tongue positioning. *The Journal of the Acoustical Society of America, 126*(4), 2033–2051.

Cos, I., Bélanger, N., and Cisek, P. (2011). The influence of predicted arm biomechanics on decision making. *Journal of Neurophysiology, 105*(6), 3022–3033.

Foulkes, P., and Docherty, G. (2006). The social life of phonetics and phonology. *Journal of Phonetics, 34*(4), 409–438.

Feldman, A. G. (1986). Once more on the equilibrium-point hypothesis (λ model) for motor control. *Journal of Motor Behavior, 18*(1), 17–54.

Flanagan, J. R., Ostry, D. J., and Feldman, A. G. (1990). Control of human jaw and multi-joint arm movements. In G.E. Hammond (Ed.), *Cerebral Control of Speech and Limb Movements* (pp. 29–58), North-Holland: Elsevier Science Publishers B.V.

Frère, J., and Hug, F. (2012). Between-subject variability of muscle synergies during a complex motor skill. *Frontiers in Computational Neuroscience*, 6, 99.

Fitch, W. T., and Giedd, J. (1999). Morphology and development of the human vocal tract: A study using magnetic resonance imaging. *The Journal of the Acoustical Society of America, 106*(3), 1511–1522.

Franklin, D. W., Liaw, G., Milner, T. E., Osu, R., Burdet, E., and Kawato, M. (2007). Endpoint stiffness of the arm is directionally tuned to instability in the environment. *The Journal of Neuroscience, 27*(29), 7705–7716.

Fuchs, S. and Perrier, P. (2005). On the complex nature of speech kinematics. *ZAS Papers in Linguistics, 42*, 137–165.

Fuchs, S., Winkler, R., and Perrier, P. (2008). Do speakers' vocal tract geometries shape their articulatory vowel space? In *Proceedings of ISSP 2008 – 8th International Seminar on Speech Production*, pp. 333–336, Univ. Strasbourg, France.

Fuchs, S., Toda, M., and Żygis, M. (2010). Do differences in male versus female /s/ reflect biological or sociophonetic factors? In Fuchs, S.,

Toda, M., and Żygis, M. (Eds.), *Turbulent sounds: An interdisciplinary guide* (pp.281–302), Walter de Gruyter.

Fuchs, S., Perrier, P., and Hartinger, M. (2011). A critical evaluation of gestural stiffness estimations in speech production based on a linear second-order model. *Journal of Speech, Language, and Hearing Research, 54*(4), 1067–1076.

Gay, T., Boë, L.-J., and Perrier, P. (1992). Acoustic and perceptual effects of changes in vocal tract constrictions for vowels. *The Journal of the Acoustical Society of America, 92*(3), 1301–1309.

Gray, H. (1918). Anatomy of the human body. Philadelphia: Lea and Febiger, in Bartleby.com, 2000.

Gribble, P. L., and Ostry, D. J. (1996). Origins of the power law relation between movement velocity and curvature: modeling the effects of muscle mechanics and limb dynamics. *Journal of Neurophysiology, 76*(5), 2853–2860.

Harshman, R., Ladefoged, P., and Goldstein, L. (1977). Factor analysis of tongue shapes. *The Journal of the Acoustical Society of America, 62*(3), 693–707.

Hazen, K. (2002). Identity and language variation in a rural community. *Language, 78*(2), 240–257.

Hill, A.V. (1938). The heat of shortening and the dynamic constants of muscle. *Proceedings of the Royal Society B: Biological Sciences, 126*,136–195.

Honda, K. (1996). Organization of tongue articulation for vowels. *Journal of Phonetics, 24*(1), 39–52.

Hug, F., Turpin, N. A., Guével, A., and Dorel, S. (2010). Is interindividual variability of EMG patterns in trained cyclists related to different muscle synergies? *Journal of Applied Physiology, 108*(6), 1727–1736.

Jackson, M. T. (1988). Analysis of tongue positions: Language-specific and cross-linguistic models. *The Journal of the Acoustical Society of America, 84*(1), 124–143.

Kelso, J. S., Vatikiotis-Bateson, E., Saltzman, E. L., and Kay, B. (1985). A qualitative dynamic analysis of reiterant speech production: Phase portraits, kinematics, and dynamic modeling. *The Journal of the Acoustical Society of America, 77*(1), 266–280.

Laboissière, R., Ostry, D. J., and Feldman, A. G. (1996). The control of multi-muscle systems: human jaw and hyoid movements. *Biological Cybernetics*, *74*(4), 373–384.

Ladefoged, P. (1984). Out of chaos comes order: Physical, biological, and structural patterns in phonetics. *Proceedings of the 10th International Congress of Phonetic Sciences*, pp. 83–95.

Ladefoged, P., and Maddieson, I. (1996). *The sounds of the world's languages*. Oxford: Blackwell.

Lindblom, B. (1990). Explaining phonetic variation: A sketch of the H&H theory. In Hardcastle, W. J., and Marchal, A. (Eds.), *Speech production and speech modelling* (pp. 403–439). Springer, The Netherlands.

Mooshammer, C., Hoole, P., and Kühnert, B. (1995). On loops. *Journal of Phonetics*, *23*(1), 3–21.

Munson, B., and Babel, M. (2007). Loose lips and silver tongues, or, projecting sexual orientation through speech. *Language and Linguistics Compass*, *1*(5), 416–449.

Nazari, M. A., Perrier, P., Chabanas, M., and Payan, Y. (2010). Simulation of dynamic orofacial movements using a constitutive law varying with muscle activation. *Computer Methods in Biomechanics and Biomedical Engineering*, *13*(4), 469–482.

Nazari, M. A., Perrier, P., Chabanas, M., and Payan, Y. (2011). Shaping by stiffening: a modeling study for lips. *Motor Control*, *15*(1), 141–168.

Payan, Y., and Perrier, P. (1997). Synthesis of VV sequences with a 2D biomechanical tongue model controlled by the Equilibrium Point Hypothesis. *Speech Communication*, *22*(2), 185–205.

Perrier, P., Boë, L.-J., and Sock, R. (1992). Vocal tract area function estimation from midsagittal dimensions with CT scans and a vocal tract cast: Modeling the transition with two sets of coefficients. *Journal of Speech and Hearing Research*, *35*, 53–67.

Perrier, P., Lœvenbruck, H., and Payan, Y. (1996). Control of tongue movements in speech: The equilibrium point hypothesis perspective. *Journal of Phonetics*, *24*(1), 53–75.

Perrier P., Perkell J. S., Payan Y., Zandipour M., Guenther F., and Khaligi A. (2000). Degrees of freedom of tongue movements in speech may be constrained by biomechanics. In *Proceedings of the 6th International*

Conference on Spoken Language and Processing (ICSLP). (Vol 2., pp. 162–165). Beijing, China.

Perrier, P., Payan, Y., Zandipour, M., and Perkell, J. (2003). Influences of tongue biomechanics on speech movements during the production of velar stop consonants: A modeling study. *The Journal of the Acoustical Society of America, 114*(3), 1582–1599.

Perrier, P., and Fuchs, S. (2008). Speed–curvature relations in speech production challenge the 1/3 power law. *Journal of Neurophysiology, 100*(3), 1171–1183.

Rogers, C. R., Mooney, M. P., Smith, T. D., Weinberg, S. M., Waller, B. M., Parr, L. A., Docherty, B. A., Bonar, C. J., Reinholt, L. E., Dleyiannis, F. W.-B., Siegel, M. I., Marazita, M. L., and Burrows, A. M. (2009). Comparative microanatomy of the orbicularis oris muscle between chimpanzees and humans: evolutionary divergence of lip function. *Journal of Anatomy*, *214*(1), 36–44.

Schwartz, J.-L., Beautemps, D., Abry, C., and Escudier, P. (1993). Inter-individual and cross-linguistic strategies for the production of the [i] vs. [y] contrast. *Journal of Phonetics, 21,* 411–425.

Shiller, D. M., Sato, M., Gracco, V. L., and Baum, S. R. (2009). Perceptual recalibration of speech sounds following speech motor learning. *The Journal of the Acoustical Society of America*, *125*(2), 1103–1113.

Stavness, I., Gick, B., Derrick, D., and Fels, S. (2012). Biomechanical modeling of English /r/ variants. *The Journal of the Acoustical Society of America*, *131*(5), EL355–EL360.

Stavness, I., Nazari, M. A., Perrier, P., Demolin, D., and Payan, Y. (2013). A biomechanical modeling study of the effects of the orbicularis oris muscle and jaw posture on lip shape. *Journal of Speech, Language, and Hearing Research*, *56*(3), 878–890.

Stavness, I., Nazari, M. A., Flynn, C., Perrier, P., Payan, Y., Lloyd, J. E., and Fels, S. (2014). Coupled biomechanical modeling of the face, jaw, skull, tongue, and hyoid bone. In *3D Multiscale Physiological Human* (pp. 253–274). Springer London.

Story, B. H. and Titze, I. R. (1995). Voice simulation with a body-cover model of the vocal folds. *The Journal of the Acoustical Society of America, 97*, 1249–1260.

Story, B. H., Laukkanen, A.-M., and Titze, I. R. (2000). Acoustic impedance of an artificially lengthened and constricted vocal tract. *Journal of Voice 14*(4), 455–469.

Takemoto, H. (2001). Morphological analyses of the human tongue musculature for three-dimensional modeling. *Journal of Speech, Language, and Hearing Research*, *44*(1), 95–107.

Titze, I. R. and Story, B. H. (2002). Rules for controlling low-dimensional vocal fold models with muscle activation. *The Journal of the Acoustical Society of America, 112*, 1064–1076.

Whalen, D. H. (1990). Coarticulation is largely planned. *Journal of Phonetics*, *18*, 3–35.

Viviani, P., and Stucchi, N. (1992). Biological movements look uniform: evidence of motor-perceptual interactions. *Journal of Experimental Psychology: Human Perception and Performance*, *18*(3), 603–623.

Winkler, R., Fuchs, S., Perrier, P., and Tiede, M. (2011a). Speaker-specific biomechanical models: From acoustic variability via articulatory variability to the variability of motor commands in selected tongue muscles. In *9th International Seminar on Speech Production (ISSP 2011)* (pp. 219–226). Montréal Canada.

Winkler, R., Fuchs, S., Perrier, P., and Tiede, M. (2011b). Biomechanical tongue models: An approach to studying inter-speaker variability. In *12th Annual Conference of the International Speech Communication Association (Interspeech 2011)* (pp. 273–276).

Winters, D. A. (2009). *Biomechanics and motor control of human movement* (4th Edition). John Wiley & Sons, Inc.

Jean-François Bonastre[1], Juliette Kahn[2], Solange Rossato[3] and Moez Ajili[1]

[1]*Laboratoire d'Informatique d'Avignon, Avignon*
[2]*Laboratoire National de Métrologie et d'Essais, Paris*
[3]*Laboratoire d'Informatique de Grenoble, Grenoble*

Forensic Speaker Recognition: Mirages and Reality

Abstract: Forensic speaker recognition is a topic similar to a tropical climate, where a big storm could form any day. It is a particularly controversial topic for three main reasons: the nature of the material it relies on, the maturity of scientific knowledge in this field, and its history. Forensic speaker recognition is under the spotlight because there is a huge and increasing demand for expertise in courts. In this chapter, the importance of the Bayesian decision framework is highlighted, which is now the standard paradigm for forensic speaker comparison, and reopens the question of science in court. The impressive progresses achieved in the field of Automatic Speaker Recognition (ASR) during the last decade are acknowledged. This raises the question of the use of ASR in forensic voice comparison. In this context we point out on several important weaknesses in the evaluation protocols, insisting on the fact that the whole communication process has to be taken into account, including speaker specificities not only from a speech production perspective but also from the perspective of the interactions with the interlocutors. The final objective is to reaffirm strongly in the scientific area the "need for caution message" concerning forensic speaker recognition applications in courts.

1. Introduction

Forensic speaker recognition is a hot topic primarily because of the forensic aspect. In the forensic field, mistakes have a direct impact on humans, on their lives. Forensics is also an area that science and law have to share. It is not easy, but also not impossible (Roberts, 2013). These two characteristics tend to make forensics similar to a tropical climate, where a big storm could form any day. Forensic speaker recognition is under the spotlight because there is a huge and increasing demand for expertise in courts. This is mainly due to the development of modern communication services: it is becoming increasingly rare to see a law case, in which there is no mention of the use of a smartphone or some other modern communication tool. In addition to sharing the 'hot' nature common to all forensic media, speaker recognition is also a particularly controversial topic for three main reasons: the nature of the material it relies on, the maturity of scientific knowledge in this field, and its history.

1.1. Should speech analysis be regarded as physical biometric?

Firstly, speech is not exclusively a physical biometric. Language is alive: if part of language is rooted in our genetic makeup, most of it is learned and varies consistently over time. The idea that we can recognize individuals by their voices is widespread among people. The main reason is that speech is a human activity, directly attributed to a human speaker. When hearing speech, one imagines a speaker, and assigns sex, age, geographical and social origin, and even some personality features. Moreover, speech is constrained by diastratic, diatopic, and diaphasic variations. Sociolinguistics studies how languages differ among social groups (because of e.g. age, sex, level of education, status, and ethnicity; see Labov, 1972), while geolinguistics is concerned with the spatial distribution of linguistic phenomena. Pragmatics studies how speech production depends not only on phonology, lexicon and syntax but also on the inferred intent of the speaker and the context of the utterance (Austin, 1970). Speech also conveys the emotional and psychological states of the speaker (Scherer, 1986). All these factors may influence the realization of a speech utterance. Voice "biometrics" aims to identify idiosyncratic features in the speech signal produced at a given time and with a given communicative intention. This task is difficult because a speech signal is not a direct reading of body traces (like fingerprints or DNA), and includes a large variability caused by factors such as speech acts, languages or speaker's roles, without even taking into account the possibility of intentional changes in voice (disguise) or speaker-independent conditions like noise. It clearly appears that voice authentication is largely based on behavioral variables: it looks at the way one is speaking, not the physical properties of her/his body! If the notion of behavioral biometrics is accepted, then speech could be considered as being related to it.

1.2. Lack on commonly accepted approaches or techniques

In addition to the difficulty related to speech not being a true (well-defined) biometric feature, there is a clear lack of scientifically accepted knowledge, approaches or techniques in the field of forensic speaker recognition. This is a consequence of the nature of the material studied — human language mediated through speech. As explained in the previous paragraph, the large number of open variables creates a real and complex difficulty in experimental assessment. Gathering experimental confirmation from a sample database with specific conditions will not allow the scientist to generalize the results to other conditions. Most often, the researchers will have to propose new hypotheses and do the experiments again with other conditions. The lack of commonly accepted methods in forensic speaker recognition is linked to this variability, as well as to the involvement of multiple scientific areas such as acoustic phonetics, signal processing, phonology and other linguistic disciplines.

1.3. Historical charlatanry and controversy in forensic speaker recognition

Finally, forensic speaker recognition is also a hot and controversial topic due to its history, with the existence of some charlatanry in the field since the sixties. In 1962, Kersta introduced the misleading term "Voiceprint identification", referring to the speech spectrogram representation. However, this is only a visual representation of the speech, based on the acoustic properties of speech which result from articulatory movements controlled by the speaker. It does not trace the speaker himself (Bolt et al., 1970). This could be seen as a classical scientific controversy of the past but this misconception still holds: several associations of forensic speaker recognition experts still remind us that speech spectrograms should not be used in their "best practices" or resolutions. For example, in 2007, the IAFPA[1] voted a resolution[2] considering that the spectrogram comparison approach (with a methodological reference to Tosi, 1979) is "without scientific foun-

1 International Association for Forensic Phonetics and Acoustics (http://www.iafpa.net/)

2 http://www.iafpa.net/voiceprintsres.htm

dation and it should not be used in forensic casework". This resolution was proposed and voted 37 years after Bolt's paper (Bolt, 1970), which clearly indicates that this misleading visual representation of speech was still used by some "experts" in 2007, despite any scientific evidence. Boë (2000) described the "Voiceprint" history in detail, as well as several other examples of science misused in forensic speaker authentication, like the Micro-Surface "REVAO" tool in France during the 1984 "Gregory" case[3] or the "Prieto" case[4]. In these cases, the methods used by the "experts" were questioned by the court and finally rejected. Unfortunately, we are not only talking about history. For instance, Morrison (2014) discussed about "distinguishing between forensic science and forensic pseudoscience". The need for reaffirming the unscientific aspects of spectrogram reading in 2007 is reinforced by recent charlatanism in different aspects of forensic speech science, as highlighted in recent articles (Eriksson and Lacerda, 2007; Boë and Bonastre, 2012).

To conclude this introduction, while the first important novelty in voice comparison area comes from the general acceptance of the Bayesian paradigm, which reopens the question of science in court, the real innovation is the strong emergence of Automatic Speaker Recognition (ASR) processes. Over the last few decades, automatic systems have improved from error rates of around 20% to error rates of less than 1%, even though the difficulty of the task has increased significantly. This raises the question of the use of ASR in forensic voice comparison. In the next sections, we will focus on these two main aspects, with a short side note on voice convergence phenomena.

2. Bayesian decision framework: Evolution or revolution?

"Would jurists accept that the concept of reasonable doubt on the identification of a suspect escapes their province and that the threshold is imposed onto the court by the scientist?" This question asked by Christophe

3 Grégory Villemin was a young boy murdered in 1984. This unresolved case involves several members of his family and is very famous in France.

4 Jérôme Prieto was accused of participating in a Basque terrorism case which took place in 1996 on the basis of a recorded phone message.

Champod and Didier Meuwly (Champod and Meuwly, 2000[5,6]) marks an important change in the understanding of "forensics" by the speaker recognition community. Champod and Meuwly's work followed several similar studies in forensics, like the one of Balding and Donnelly (1994) for DNA, but it was the first for forensic voice comparison. While automatic speaker recognition researchers were working on how to decrease the probability of a false identification in a forensic report, they were debating hotly in order to know if this probability is well-known and low enough to authorize forensic applications[7]. Champod and Meuwly showed to all experts that they are not in charge of making decisions. They have to provide the court with an evaluation which illustrates the convincing force of the results, not to take part in the judicial debate. This is definitively not possible if science says: "the suspect is guilty" through scientific expertise. Unfortunately, in many trials, saying that a suspect is the one speaking in a given trace/recording is actually equivalent to stating that he or she is guilty.

In other words, with the Bayesian paradigm, the speech scientist does not "identify" people, but provides the jury with the specialist's scientific information in a procedure which is conceptually identical to the one used nowadays in the presentation of DNA evidence.

2.1. Implementation of the Bayesian decision framework in forensic trials

Scientifically speaking, the need for the expert to stay out of the province of jurists is implemented using the Bayesian decision framework. Based on a piece of evidence E (a vocal message X), the experts have to present their conclusion using a Likelihood Ratio (LR), which expresses how likely the evidence is under the prosecutor's hypothesis (the suspect pronounced message X), versus the defender's hypothesis (the suspect did not pronounce

5 Firstly presented in Christophe Champod's tutorial, RLA2C, Avignon, 1998 (RLA2C was one of the precursors of "Speaker Odyssey" workshops).

6 As presented by the authors, this sentence was inspired by the report of a panel on statistical assessments as evidence in courts (Fienberg, 1989, p. 141), from which the following quotation is taken "it is the utility function of the court that is appropriate, not the utility function of the statistician".

7 Of course, these questions were important and still are. We will get back to these aspects later, in the light of the Bayesian decision framework.

message *X*). The *LR* is presented in equation 1, where H_1 is the prosecutor's hypothesis and H_2 is the defender's hypothesis (This formula differs slightly from Champod (2000) and will be explained later):

$$LR = \frac{\Pr(E|H_1)}{\Pr(E|H_2)} \qquad (1)$$

The numerator is the probability of the evidence given H_1 and the denominator is the probability of the evidence given H_2. While the numerator can be estimated by the expert by considering the evidence and the suspect, the denominator is the random match probability, "which can be derived from an objective or subjective estimation of the relative frequency of the concordant features in the relevant population" (Champod and Meuwly, 2000).

The use of the *LR* framework for the forensic expert's report is very attractive. As expected, it places the expert on neutral ground by withdrawing the need for her/him to conclude the report with a "decision". Furthermore, it also helps the expert to follow a scientific approach, since work is based only on evidence *E*.

However, the *LR* alone is not sufficient for the court, which must also consider the posterior odds of the two hypotheses H_1 and H_2, as expressed in equation 2:

$$\frac{\Pr(H_1|E,I)}{\Pr(H_2|E,I)} = \frac{\Pr(E|H_1)}{\Pr(E|H_2)} \times \frac{\Pr(H_1|I)}{\Pr(H_2|I)} \qquad (2)$$

In equation 2, the LR issued by the expert can be recognized. This LR is multiplied by the ratio of the prior probabilities of H_1 and H_2, respectively, $\Pr(H_1|I)$ and $\Pr(H_2|I)$. These prior probabilities are based on all the elements of the case, denoted *I* here. They may change during the law case or the trial, for example due to new elements added in *I*. Paraphrasing Champod and Meuwly, the prior probabilities are clearly in the province of the jurist and the court, and not in the province of the expert.

Although this Bayesian formalism was new for most caseworkers engaged in forensic speech comparison in 1998/2000, it is now widely accepted and considered by many experts as the logically correct framework (Rose, 2006; Gonzalez-Rodriguez et al., 2007; Jessen, 2008). It is interesting to notice that the references provided come mainly from the articulatory-phonetic voice comparison community. It clearly shows the wide acceptance of the (Bayesian) likelihood ratio for forensic voice comparison, even if this question is still debated, mainly when discussing what should be presented in courts (French and Harrison, 2007; Rose and Morrison, 2009; French et al., 2010). Gold and Hughes (2014) presented recently an interesting survey on the use of "numerical likelihood ratio framework to forensic speaker comparison", which emphasizes both the advantages of the *LR* approach and its practical difficulties.

2.2. Bayesian decision framework limitations for forensic trials

As shown previously, the Bayesian formalism becomes a cornerstone of forensic expertise and is reported in several areas, including speech. It provides a very elegant theoretical framework and places the expert (back) in her/his proper domain which is science and not judgment. However, implementing a theoretical framework to handle real-world cases requires some "adaptations" and causes three main problems:

a. Estimation of $Pr(E|H_2)$

$Pr(E|H_2)$ plays a very important role in *LR*, at least equivalent to $Pr(E|H_1)$, even if the former is clearly underrepresented in the voice comparison and speaker recognition literature. Estimating the probability is not an easy task. For example, with a machine learning approach, it is possible to learn a class model for H_1 using several samples of the suspect's voice while it is not trivial to train such a model for H_2 as it is more difficult to find samples of a "non-voice". H_2 implies that the speech recording under scrutiny was pronounced by someone else other than the suspect. Hence the corresponding class represents all voices except the suspect's.

In Champod and Meuwly (2000), $Pr(E|H_2)$ is the "random match probability" and "can be derived from an objective or subjective estimation of the relative frequency of the concordant features in the relevant population". It is interesting to read that for H_2, the notion of "subjective evaluation" is introduced in the scientific process of the forensic expert.

Furthermore, it is important to notice that three elements have to be evaluated in order to estimate $Pr(E|H_2)$: the **concordant features**, their **relative frequency** and the **relevant population**. This means that a forensic approach claiming to comply with the Bayesian formalism, which is very often described as the only scientific formalism accepted for forensic evidence, should define these three elements explicitly. And the latter does not depend on the forensic expert, or at least not completely, since it is "dictated by the hypothesis proposed by the defense" (Champod and Meuwly, 2000). It means that the forensic expert referral should include a clear description of the expected relevant population. We should also remember that this hypothesis is not definitive and may evolve during the trial.

b. Background information

Frequently, the forensic expert has access to several pieces of background information concerning the current case, other than the piece of evidence *E* and the hypotheses to be evaluated. Therefore, the *LR* equation very often includes *I'*, a subset of *I*, in addition to the evidence *E* in the expert knowledge:

$$LR = \frac{\Pr(E|H_1,I')}{\Pr(E|H_2,I')}$$

Forensic experts often have an unrestricted access to the background information. Consequently, the *LR* is often formulated using *I'* = *I* (Champod and Meuwly, 2000).

We saw earlier that the *LR* denominator is an estimation of the random probability in the "relevant population". This is understandable if the expert wishes to use as much information as possible in order to determine the H_2 probability. Unfortunately, this is in obvious contradiction with the scientific position, which is to be as little subjective as possible. It may be useful here to remember the well-known double blind principle and why it is so important in medical research assessment.

So, the question is: Could a completely scientific and objective assessment be achieved if the expert has additional information beyond the evidence itself, for example about the suspect's origins, preferences and criminal record? If the answer is no and if we want to keep expert's reports as scientific as possible, it is important to clearly define which information can be provided to the experts. More generally speaking, this problem is known as the "forensic confirmation bias". The clearest example of this bias is given by the high-profile mistaken fingerprint identification of Brandon Mayfield in the Madrid Bomber case (Kassin et al., 2013). From a juristic point of view, it also seems important to make sure that the details provided to the experts are accessible, case by case, to the various parties, e.g. the defender's.

c. Understandability of *LR* by the court

Champod and Meuwly (2000) claim that *LR* is useful "for assisting scientists to assess the value of scientific evidence" and to "clarify the respective roles of scientists and of members of the court". These two claims have been discussed previously and are quite easy to accept. But Champod and Meuwly also claim that *LR* is useful to "help jurists to interpret scientific evidence". Of course, a forensic analysis has an interest only if judges, lawyers, and jurors are able to understand the work done by the expert precisely, as well as the intrinsic nature of the scientific evidence presented.

However, understanding probabilities in general and *LR* more specifically is not straightforward. Daniel Kahneman, the 2002 economics Nobel Prize (co)laureate, a specialist of judgement and decision-making and one of the two proposers of the prospect theory (Tversky and Kahneman, 1974), states in his 2011 book "Thinking, Fast and Slow" that Bayesian reasoning is not natural for humans. This is not only true for normal people but also for statistics specialists. In Thompson et al. (2013), the perception of *LR* by jurors is analyzed. It appears that it is not easy for them to correctly understand statistical evidence. As highlighted by the authors, this is particularly true when forensic experts, prosecutors or lawyers provide arguments that invite or encourage fallacious conclusions from statistical evidence, which is not uncommon in courts.

Moreover, as Bayesian theory as well as statistics and probabilities in general are now a mandatory part of forensic evidence presentation and understanding, it would be interesting to include serious courses in these areas in law studies curricula, which is not often the case at present.

3. Automatic approaches: a new avenue for forensic speaker recognition?

The use of automatic approaches for forensic speaker recognition clearly offers important advantages, in terms of objectivity and repeatability of the voice comparison measures but also in terms of human time costs. The limited cost of automatic processes also can allow the expert to test several voices against the piece of evidence, which is a clear progress towards double blind, objective procedures. This interest in the use of automatic systems for forensic applications has been present in the literature for a long time (Nakasone and Beck, 2001; Alexander et al., 2005; Drygajlo, 2007; Becker et al., 2010; Mandasari et al., 2011).

For decades, from the early ages of speaker recognition (Pruzansky, 1963) until the end of the past millennium, the performance of automatic speaker recognition systems were so poor that using it for real forensic cases was not feasible yet. The situation began to change with new statistics-based approaches and the large scale speaker recognition evaluation campaigns organized by the NIST since 1996 (Przybocki and Martin, 2004). In order to take this evolution into account, several scientific institutions (see Bonastre et al., 2003) sent a clear need-for-caution message concerning the use of automatic speaker recognition technologies and for forensic speaker authentication in general to the forensic field, including statements such as, "currently, it is not possible to completely determine whether the similarity between two recordings is due to the speaker or to other factors", "caution and judgment must be exercised when applying speaker recognition techniques, whether human or automatic" or "at the present time, there is no scientific process that enables one to uniquely characterize a person's voice or to identify with absolute certainty an individual from his or her voice.".

Campbell et al. (2009) started from this "need for caution" message and revisited it in light of the impressive improvement in terms of (measured) performance made during the last decade in the field of automatic speaker recognition (see Przybocki et al., 2006, 2007; Fauve et al., 2007). They observed that the performance measured in terms of Equal Error Rates (EERs) dropped from around 9% for the year 2000 system (Reynolds et al., 2000; Bimbot et al., 2004) to 4.5% for the 2006/2007 system (Kenny et al., 2007). The EER even goes down as far as about 1% when longer training excerpts or unsupervised speaker adaptation are used (Barras et al., 2004; McLaren et al., 2008, 2011). Since 2009, the progress in terms of error rate decrease is still noticeable, mainly thanks to the "iVector" approach (Kenny et al., 2007; Bousquet et al., 2014). Nowadays, EERs lower than 1% are obtained on quite large scale evaluation sets, with millions of voice comparisons. Figure 1 proposes a schematic view of the evolution of EER over the last 2.5 decades. It is indeed a schematic view, since experimental protocols evolved over the years and are not directly comparable.

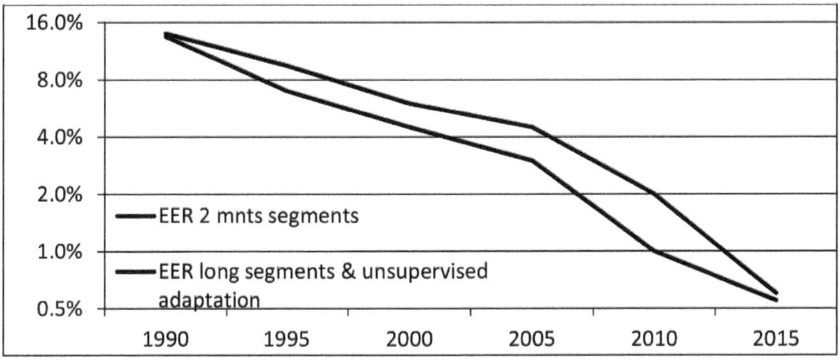

Figure 1: Schematic view of speaker detection error rates.

Recently, several studies investigated the use of Deep Neural Networks for automatic speaker recognition (Stafylakis et al., 2012; Lei et al., 2014; Kenny et al., 2014; Vasilakakis et al., 2013). The presented results show clearly that this approach is able – or will be soon – to bring an additional and significant decrease of the error rates.

While reporting these impressive progresses and error rates, it is interesting to question the results of these studies. In Campbell et al. (2009) the authors showed that an error rate is often not enough to understand the behavior of a system. In the following paragraphs, we propose a fresh look at the performance related numbers, their meaning and their limits.

3.1. Instability, imprecision and inadequacy of the performance measures

An unquestionable advantage of automatic approaches for forensic applications is to offer the ability to assess the techniques on a large number of voice comparison trials. For example, in NIST SRE evaluations, hundreds of thousands of tests are done. The impressive error rates reported earlier in this chapter are obtained with this kind of experimental protocols. The robustness of such an evaluation protocol relies on respecting some straightforward rules (Phillips et al., 2000; Petrovska et al., 2009) and on "brute force", i.e. the size of the evaluation set. Particularly in the NIST

SRE evaluation, when a system is working on a voice comparison between recordings *X* and *Y*, only the use of these two recordings is allowed in the evaluation set (i. e. knowledge of *Z* is not allowed, if *Z* is another recording of the evaluation set).

Soong et al. (1987) is one of the first speaker recognition studies showing a strong evaluation protocol: 50 male and 50 female speakers were recorded, each of the speakers pronounced 200 digits in 5 recording sessions, which corresponds to the maximum available computing power at that time. For the main testing condition, NIST SRE 2010 (NIST, 2010) involved 6 000 speaker models, 25 000 test segments and up to 750 000 voice comparison tests (for one testing condition). Looking at the magnitude factor between the two experiments reported here, it is easy to understand why there has been a small interest in evaluation protocols in the last decades: the progress made by the computers, following Moore's law and reflected by the size of the databases, gave a strong impression of increasing robustness, based on the "brute force" aspect alone.

During that period, performance was measured only by using global error rates averaged on the whole test set[8]. This way of evaluating the performance of speaker recognition systems presents two main drawbacks: the criterion itself and the global nature of the performance measure.

The classical speaker detection performance criteria – false alarm, false reject and cost functions – depend on a decision making (on a threshold) while the Bayesian decision paradigm rejects this notion of decision for a forensic voice comparison. In the Bayesian paradigm, the systems output – a likelihood ratio. Its value – is meaningful in itself, not simply because this *LR* is large enough (or small enough) to allow a "good" decision compared to a threshold. For example, we expect *LRs* with a low power (close to 1) when the piece of evidence contains little speech material, i. e. little speaker-specific information. The same effect is expected if the quality of the audio material is low. This is well described in Morrison (2011). The authors use the notions of *validity/accuracy* and *precision*, which are illustrated in Figure 2. If this approach is clearly the accepted one, we would like a solution which is able to represent both notions in one number. The

8 Mainly false alarm, miss probability, EER, DCF and DET plots (Martin et al., 1997)

"log-likelihood ratio cost function" introduced by Brümmer (Brümmer and du Preez, 2006; van Leeuwen and Brümmer, 2013) denotes C_{LLR}, a value that could be seen as the best available solution to the problem. C_{LLR} is an *LR*-oriented performance criterion based on assumptions about *LR* distributions. Although some of its underlying hypotheses are not always validated in practice, C_{LLR} is now the official criterion for NIST Speaker Recognition Evaluations and Language Recognition Evaluations. C_{LLR} also allows to separate *calibration loss* and *discrimination loss*. *Calibration loss*[9] is a loss due to badly formatted *LR* values, a problem which could be solved with an adequate calibration process ("calibration" is often used as another word for "normalization"). *Discrimination loss* corresponds to the rest of the losses, which comes from the two speech recordings and from the system itself.

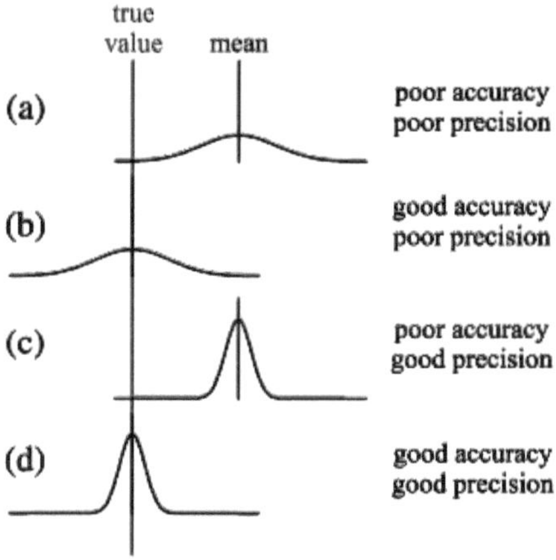

Figure 2: Schematic view of accuracy and precision (Morrison, 2011)

The second drawback with some evaluation performance measures is the way evaluation data is used or the way the evaluation database and protocol

9 An example could help us to define "calibration loss". Let us imagine that we have a perfect system which outputs perfect LRs. Now, something like a constant background noise disturbs this system and adds a constant bias to its output. Of course, the C_{LLR} of the system will improve significantly while its discrimination power is still the same. The difference between the two C_{LLR} is "calibration loss".

were designed. Until now, a test condition has been defined only by little information, such as the durations of the two speech files composing a voice comparison, the language used and the "channel" (e.g. close or distant microphone, fixed phone or cellphone). All the available voice comparison samples corresponding to these conditions are taken together and a global performance is computed in terms of classical error rates or C_{LLR}. The robustness of the evaluation for the given test condition once again relies on "brute force": a large number of voice comparison samples[10]. The number of samples per speaker and the characteristics of the speaker are not taken into account, except the sex and the mother tongue of the speaker. It is amazing to observe that the "speaker factor" is still not taken into account in the design of evaluation plans even though its great influence is well-known. Doddington et al. (1998) showed that, for an automatic speaker recognition system, there are different "speaker profiles". Depending on their "profile", only a few speakers are responsible for a large part of the errors reported. The authors showed that the performance measures significantly depend on this factor.

Revisiting the perspective opened by Doddington et al. (1998), Kahn et al. (2010) demonstrated that the notion of "speaker profile" is in fact a simplified view of a more general problem: speaker recognition systems model speech files and not, or not only, the speech or the voice of a given speaker. In order to demonstrate this assumption, the authors built a new experimental setup using the NIST 2008 evaluation database. The experiment was composed of voice comparison trials, represented by a couple of speech signals (X^i, Y_k). The right value, Y_k, is fixed and simply one of the K speech extracts from recording set Y. The left value, X^i, is the in-interest factor. X^i is the recording of speaker S^i, taken from a subset of recordings X^i_j, pronounced by S^i. For each S^i speaker, voice comparison trials (X^i, Y_k) with k varying from 1 to K are carried out using each available speech signal X^i_j, j varying from 1 to J. For each S^i speaker, the speech extract which allowed the speaker recognition system to make the least errors is labelled with a "best" label. Conversely, the speech extract showing the maximum number of errors is labelled with a "worst" label. Figure 3 plots

10 The number of different speakers involved in the condition is often taken into consideration.

the performance of the system when the recordings selected for the X^i parts of the voice comparisons are the "best" ones or the "worst" ones. The EER moves from less than 5% for the recordings with the "best" labels to more than 20% with the "worst" labels[11]. It is important to emphasize that the only difference between the "best" condition and the "worst" condition is the speech sample selected to represent a given speaker[12]. Clearly, the speaker recognition system gives a great importance to the speech extract itself. In forensic voice comparison, it means that the choice of the speech material used as comparison has an important effect on the voice comparison result itself.

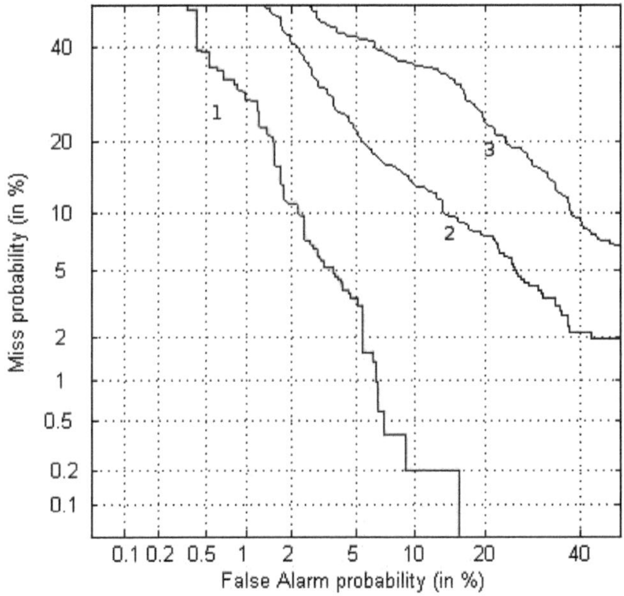

Figure 3: DET performance curves of a speaker recognition system using (1) the "best" speech extracts, (3) the "worst" speech extracts and (2) randomly selected speech extracts (Kahn et al., 2010).

11 Kahn et al.(2010) reported similar performance differences when different databases or systems are used.

12 All the speech excerpts are coming from the same evaluation condition of NIST 2008, in order to limit biases like channel, language or duration.

Even if we have just highlighted the constraints on speech recognition systems, it is important to remember that international evaluations like NIST SRE and HASR (Greenberg et al., 2010, 2011; Martin et al., 2014), NFI-TNO (van Leeuwen et al., 2006) or AHUMADA (Ortega-Garcia et al., 2000) have allowed us to discover or evaluate several variability factors over the years.

HASR is an interesting and specific case as it merges phonetic-forensic aspects with the aspects of automatic approaches. NIST HASR initiative started in 2010. It is based on a short subset of trials extracted from the NIST-SRE evaluation set[13]. The trials are processed by human experts who are allowed to use automatic tools. This initiative was at the origin of numerous studies (Schwartz, 2010; Ramos et al., 2011; Audibert et al., 2010; Shen et al., 2011; Kahn et al., 2011) or, more recently, (Hautamäki et al., 2013; van Dijk et al., 2013; Univaso et al., 2013).

Campbell et al. (2009) show a striking "voice aging" effect detected by NIST after the SRE 2005 evaluation: performance decreased significantly when the two recordings of the voice comparison trial were separated only by a few weeks. Figure 4 presents the corresponding DET curves. During his Speaker Odyssey 2014 keynote talk (Campbell, 2014), Campbell presented two other factors of variability with a potentially strong impact for forensic speaker recognition: the recording device and the microphone distance. The diversity of recording devices and the mismatched conditions for different recording devices are known problems in speaker recognition. Figure 5 shows a wide performance gap depending on the used recording device. This gap widens significantly when different devices are used for the two recordings (mismatched conditions). In Figure 6, the variability factor is the distance to the microphone. The experimental results presented, extracted from NIST SRE 2008, show an EER varying from about 1% to about 3% (a threefold difference) depending on this factor.

13 The trials were selected in 2010 depending on their "intrinsic difficulty", estimated by an automatic system. This choice could be questioned by several other variants like average difficulty selection, random selection or auditory-based selection.

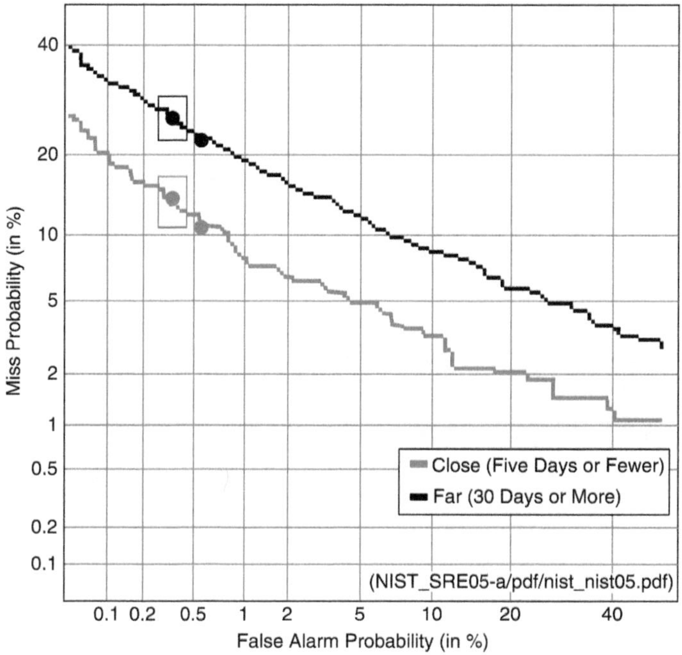

Figure 4: Performance difference reported by NIST in NIST SRE 2005, depending on the time elapsed between the recording sessions (NIST 2005 speaker recognition evaluation final meeting).

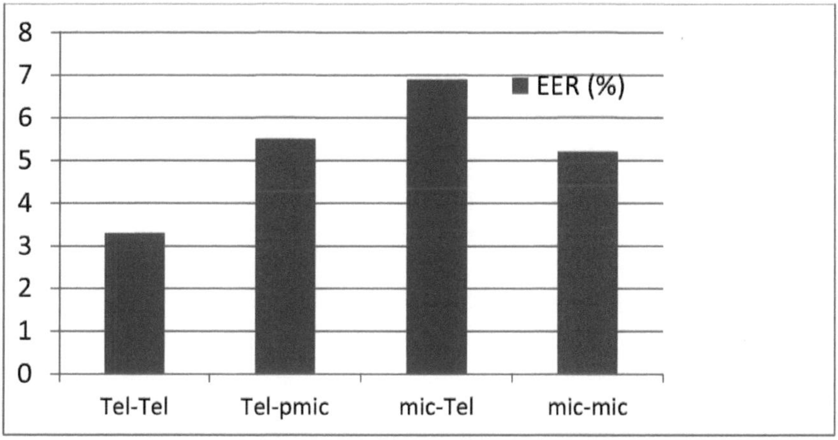

Figure 5: Effect on performance of the recording device and of mismatched recording conditions (Campbell, 2014).

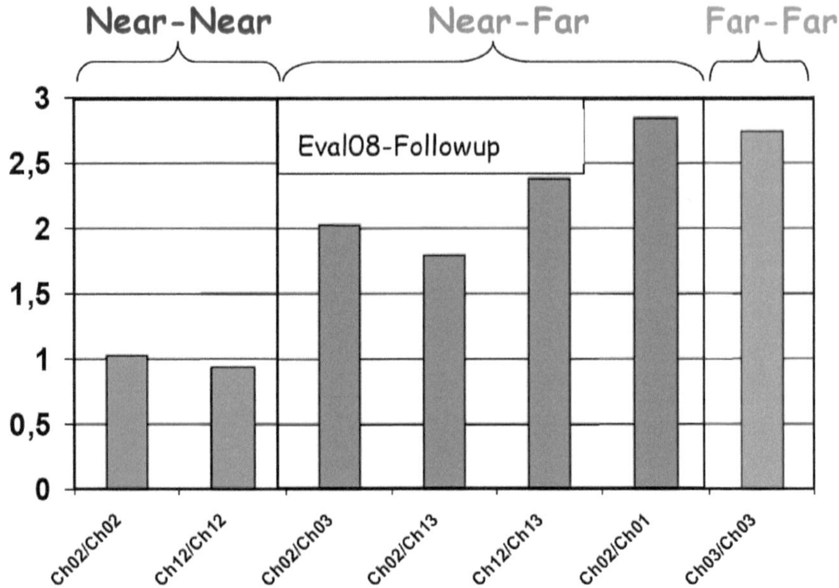

Figure 6: EER variations depending on the microphone distance (Campbell, 2014).

3.2. The speaker-specific information used by automatic speaker recognition systems

Results for the experiment in Kahn et al.(2010) are summarized in Figure 3. They show that it is not straightforward to know what information is used by automatic speaker recognition systems, even if the evaluated performances of these systems are high. Some other research work emphasizes this question. In Matrouf et al. (2006) and Bonastre et al. (2007) an artificial transformation[14] of the voice was proposed in order to spoof a speaker recognition system: after the voice transformation, the system should recognize an impostor's voice as coming from a targeted speaker. Note that only the automatic system was targeted in this spoofing experiment, not a human listener. In addition, the voice transformation should not be detected by a human listener. The targeted speaker was only described

14 The transformation is done acoustically frame by frame, only on the filter parameters of the classical source-filter model.

by a short speech sample of his/her voice (less than 2 minutes of speech), taken outside the evaluation dataset. The transformation was applied onto all the impostor trials of NIST SRE 2006 (restricted to the male trials). Table 1 reports the results of this experiment: the false alarm rate increases from 0.8% to 49.72%.

Table 1: Effect of artefact-free artificial voice transformation of impostor voices (Bonastre et al., 2007)

	False Alarm (%)	Miss probability (%)
Baseline (without transformation)	0.8	27.45
Using impostor voice transformation	49.72	27.45

The ability of this non-audible transparent transformation technique to disrupt the speaker recognition system clearly questions the nature of the information used by the system. Several researchers (Perrot et al., 2007; Zhang and Tan, 2008; Alegre et al., 2012; Wu et al., 2012; Evans et al., 2014) have done similar experiments and explored other spoofing attacks (and countermeasures), with similar comments and conclusions (Matrouf et al., 2006; Bonastre et al., 2007).

4. Voice convergence: A fundamental open question for forensic voice comparison

Quite recently, several interesting research studies have focused on voice convergence, when the interlocutors are known to establish a common ground and to align their linguistic production (Krauss and Pardo, 2006; Pardo, 2006; Babel, 2010; Kim et al., 2011). This phenomenon of interlocutor adjustment increases perceived similarity. Several acoustic attributes have been examined, such as speech rate, voice quality, formants or MFCC (Giles et al., 1991; Levitan and Hirschberg, 2011; Lelong and Bailly, 2012; Pardo et al., 2012, Pardo, 2013). This question potentially appears as a major threat against forensic speaker comparison for two reasons. First, voice convergence is an additional variability factor. Secondly, due to this type of speaker adjustments, the voice of speaker *X* could appear closer to the voice of speaker *Y* only due to the fact that *X* and *Y* participate in a conversation with the same other speaker *Z*. And to date, no scientific work excludes the hypothesis that the effects of voice convergence could remain after the conversation itself.

5. Concluding remarks

In this chapter, we firstly reminded readers of the controversial aspects of forensic speaker comparison, mainly because of the intrinsic nature of the voice, which is very different from physical biometrics like DNA or fingerprints. We highlighted the importance of the Bayesian decision framework, which has become the standard paradigm for forensics in general and for forensic speaker comparison specifically. We went deeper into the question of the use of automatic systems in forensic applications. We acknowledged the impressive progresses achieved in the field of automatic speaker recognition during the last decade, but we also pointed out several important weaknesses in the evaluation protocols. We then went back to the speaker-specific nature of the information used by automatic systems. Clearly, some doubts about automatic systems remain as demonstrated for instance by Kahn et al. (2010) and Bonastre et al. (2007). It is particularly true if we use the broader perspective of "dependability" (Avizienis et al., 2004), which takes the whole process into account. It is important to have a comprehensive picture of forensic speaker recognition processes. Campbell et al. (2009) reported the importance of calibration and Bousquet et al. (2014) showed that normalization in the iVector domain also plays a major role for the performance of a system, although there is still no theoretical explanation for this.

Previous findings on automatic speaker recognition should not give the reader the wrong impression about the use of automatic approaches in forensic speaker recognition versus human-based approaches. If automatic approaches present some weaknesses, they are unavoidable in order to assert the scientific nature of forensic speaker comparison. We do not know whether it will be possible in the future to propose a fully automatic system for forensic speaker recognition, which would follow strong scientific guidelines like Daubert's rules[15]. But we think it is quite impossible to

15 See Daubert v. Merrell Dow Pharmaceuticals, Inc. (1993) 509 U.S. 579, 589 and USA supreme court rule 702 as amended Apr. 17, 2000, eff. Dec. 1, 2000; Apr. 26, 2011, eff. Dec. 1, 2011.

meet such scientific rules without automatic processes as the typicality of each speaker-specific criterion[16] has to be assessed on very large databases. Emerging studies on tools and methods for computer-assisted approaches, like the SPAAT tool[17] used by USSS-MITLL during their HASR 2010 participation (Schwartz, 2010), demonstrate the interest of such an approach.

However, our intention is not to dismiss human expert knowledge and manual approaches. Once again, Daubert's case offers a nice proposal: "If scientific rules are not fulfilled, a witness qualified as an expert by knowledge, skill, experience, training, or education, may testify in the form of an opinion". We fully support this statement and we wish to emphasize this distinction between an expert's opinion and a scientifically-assessed method.

Finally, it is interesting to see that Campbell et al.'s (2009) conclusions are quite close to the 2003 conclusions. The main one concerns the "caution" message: "**Looking at the different points highlighted in this article, we affirm that forensic applications of speaker recognition should still be taken under a necessary need for caution. Disseminating this message remains one of the most important responsibilities of speaker recognition researchers.**" Since 2009, the research and remarks reported in this chapter have tended to significantly reinforce these conclusions.

Moving towards scientifically-sound speaker comparison approaches requires continuous research efforts. We are contributing to this effort with the work carried out within the scope of Juliette Kahn's PhD thesis for example (Kahn, 2011[18]). Figure 7 presents the logic of the work done. It reports an experiment where the part of inter-speaker variability explained by the different formants of vowels was extracted for male and female speakers. For example, the numbers reported in the figure mean that the first formant for vowel /a/ explains 15% of the inter-speaker variability. Speaker-specific information is not equally distributed on vowels and relies on the vocalic quality of sounds. These interactions between speaker-specific variability and acoustic-phonetic classes are the subject of some rare studies like (Bonastre and Meloni, 1994; Besacier et al., 2000). Further research is needed

16 A speaker specific criterion could be based on a manual or computer assisted measure on the signal.
17 Super Phonetic Annotation and Analysis Tool
18 Speech of speakers: Performance and reliability in voice biometrics

in order to provide an objective estimation of "the relative frequency of the concordant features in the relevant population" (Champod and Meuwly, 2000). Therefore, voice comparison reliability not only depends on relative frequency of the features but also on the concordance or homogeneity of the speaker-specific information classes in both speech excerpts. This work is carried on in the context of Moez Ajili's ongoing PhD[19]. Ajili (Ajili, 2015) is presenting a first measure of the data homogeneity between the two speech extracts of a voice comparison trial.

$\eta 2$		/ɐ/	/ɛ/	/o/	/e/	/ø/	/i/	/œ/	/ɔ/	/u/	/y/
H	F1	15.1%	9.7%	28.6%	7.8%	27.8%	7.4%	27.8%	21.9%	9.9%	6.9%
	F2	6.2%	17.1%	32.4%	26.7%	13.2%	13.7%	14.1%	9.9%	4.4%	10.7%
	F3	41.8%	37.0%	28.8%	24.0%	26.4%	14.7%	41.0%	30.0%	19.3%	16.1%
	F4	42.5%	38.7%	21.6%	37.9%	24.9%	19.4%	43.7%	30.2%	14.4%	20.3%
	$\eta 2$ multi varié	30%	30%	25%	29%	23%	16%	30%	26%	13%	15%
F	F1	17.5%	11.0%	11.4%	12.2%	14.2%	9.9%	28.1%	14.4%	9.8%	6.5%
	F2	12.2%	20.4%	4.7%	24.3%	7.9%	26.8%	25.6%	6.4%	4.4%	9.4%
	F3	37.4%	39.0%	43.4%	37.7%	33.6%	14.6%	52.3%	46.6%	8.8%	9.5%
	F4	37.2%	41.3%	37.6%	36.1%	25.5%	12.6%	41.6%	37.0%	17.8%	12.1%
	$\eta 2$ multi varié	28%	29%	27%	26%	21%	15%	37%	28%	11%	10%

Figure 7: Part of inter-speaker variability explained by formant and vowel. Results are given for males (H) and females (F) (Kahn, 2011).

Acknowledgments

If opinions, interpretations, conclusions, and recommendations are those of the authors, this work would not be possible without the invaluable help from Joseph P. ("Joe") Campbell and Anders Eriksson. This work was also stimulated by discussions with Reva Schwartz, Driss Matrouf, Pierre-Michel Bousquet and Guillaume Galou.

References

Aitken C.G.G., and Taroni F. (2004). *Statistics and the evaluation of evidence for forensic scientists.* 2nd ed. Wiley: Chichester.

Ajili, M., Bonastre, J.-F., Rossato, S., Kahn, J., and Lapidot. (accepted). An information theory based data-homogeneity measure for voice comparison. In *Interspeech 2015*. Dresden.

19 Moez Ajili's PhD is funded by the French National Agency funded project "Fabiole", about reliability in voice comparison.

Alegre, F., Vipperla, R., Evans, N., and Fauve, B. (2012). On the vulnerability of automatic speaker recognition to spoofing attacks with artificial signals. In *Proceedings of the 20th European Signal Processing Conference*, 36–40.

Alexander, A., Dessimoz, D., Botti, F., and Drygajlo, A. (2005). Aural and automatic forensic speaker recognition in mismatched conditions. *International Journal of Speech Language and the Law*, *12*(2), 214.

Audibert, N., Larcher, A., Lévy, C., Kahn, J., Rossato, S., Matrouf, D., and Bonastre, J. F. (2010). LIA human-based system description for NIST HASR 2010. In *Proceedings of NIST 2010 Speaker Recognition Evaluation Workshop*, Brno.

Austin, J. L. (1970). *Quand dire, c'est faire*. Seuil: Paris.

Avizienis, A., Laprie, J.-C., Randell, B., and Landwehr, C. (2004). Basic concepts and taxonomy of dependable and secure computing. *IEEE Transactions on Dependable and Secure Computing*, *1*(1), 11–33.

Babel, M. (2010). Dialect divergence and convergence in New Zealand English. *Language in Society*, *39*(4), 437–456.

Balding, D. J., and Donnelly, P. (1994). The prosecutor's fallacy and DNA evidence. *Criminal Law Review*, 711–721.

Barras, C., Meignier, S., and Gauvain, J.-L. (2004). Unsupervised online adaptation for speaker verification over the telephone. In *Odyssey 2004 – The Speaker and Language Recognition Workshop*.

Becker, T., Jessen, M., Alsbach, S., Broß, F., and Meier, T. (2010). SPES: The BKA forensic automatic voice comparison system. In *Odyssey 2010 – The Speaker and Language Recognition Workshop*.

Besacier, L., Bonastre, J.-F., and Fredouille, C. (2000). Localization and selection of speaker-specific information with statistical modeling. *Speech Communication*, *31*(2–3), 89–106.

Bimbot, F., Bonastre, J.-F., Fredouille, C., Gravier, G., Magrin-Chagnolleau, I., Meignier, S., and Reynolds, D. A. (2004). A tutorial on text-independent speaker verification. *EURASIP Journal on Advances in Signal Processing*, *4*, 430–451.

Boë, L.-J. (2000). Forensic voice identification in France. *Speech Communication*, *31*(2), 205–224.

Boë, L.-J., and Bonastre, J.-F. (2012). L'identification du locuteur: 20 ans de témoignage dans les cours de Justice. In *JEP-TALN-RECITAL Grenoble, 1*, 417–424.

Bolt, R. H., Cooper, F. S., Jr, E. E. D., Denes, P. B., Pickett, J. M., and Stevens, K. N. (1970). Speaker identification by speech spectrograms: A scientists' view of its reliability for legal purposes. *The Journal of the Acoustical Society of America*, *47*(2B), 597–612.

Bonastre, J.-F., and Meloni, H. (1994). Inter- and intra-speaker variability of French phonemes. Advantages of an explicit knowledge-based approach. In *Proceedings of the ESCA Workshop on Speaker Recognition, Identification and Verification,* Martigny, 157–160.

Bonastre, J.-F., Bimbot, F., Boë, L.-J., Campbell, J., Reynolds, D., and Magrin-Chagnolleau, I. (2003). Person authentification by voice: A need for caution. In *Proceedings of EUROSPEECH.*

Bonastre, J.-F., Matrouf, D., and Fredouille, C. (2007). Artificial impostor voice transformation effects on false acceptance rates. In *Proceedings of Interspeech, Antwerp,* 2053–2056.

Bousquet, P.-M., Bonastre, J.-F., and Matrouf, D. (2014). Exploring some limits of Gaussian PLDA modeling for i-vector distributions. In *Odyssey 2014 – The Speaker and Language Recognition Workshop*.

Brümmer, N., and du Preez, J. (2006). Application-independent evaluation of speaker detection. *Computer Speech & Language*, *20*(2–3), 230–275.

Campbell, J. P. (2014). *Speaker recognition for forensic applications. Presentation at Odyssey 2014 – The Speaker and Language Recognition Workshop*.

Campbell, J. P., Shen, W., Campbell, W. M., Schwartz, R., Bonastre, J.-F., and Matrouf, D. (2009). Forensic speaker recognition. *Signal Processing Magazine, IEEE*, *26*(2), 95–103.

Champod, C., and Meuwly, D. (2000). The inference of identity in forensic speaker recognition. *Speech Communication*, *31*(2–3), 193–203.

Doddington, G., Liggett, W., Martin, A., Przybocki, M., and Reynolds, D. (1998). Sheep, goats, lambs and wolves: A statistical analysis of speaker performance in the NIST 1998 speaker recognition evaluation. In *Proceedings of ICSLP-1998*, Sydney.

Drygajlo, A. (2007). Forensic automatic speaker recognition [Exploratory DSP]. *IEEE Signal Processing Magazine*, *24*(2), 132–135.

Gold, E., and Hughes, V. (2014). Issues and opportunities: The application of the numerical likelihood ratio framework to forensic speaker comparison. *Science and Justice, 54*(4), 292–299.

Eriksson, A., and Lacerda, F. (2007). Charlatanry in for forensic speech science: A problem to be taken seriously. *International Journal of Speech Language and the Law*, *14*(2), 169–193.

Evans, N., Kinnunen, T., Yamagishi, J., Wu, Z., Alegre, F., and Leon, P. D. (2014). Speaker recognition anti-spoofing. In S. Marcel, M. S. Nixon, and S. Z. Li (ed.) *Handbook of Biometric Anti-Spoofing* (pp. 125–146). Springer: London.

Fauve, B. G. B., Matrouf, D., Scheffer, N., Bonastre, J.-F., and Mason, J. S. D. (2007). State-of-the-art performance in text-independent speaker verification through open-source software. *IEEE Transactions on Audio, Speech and Language Processing*, *15*(7), 1960–1968.

Fienberg, S. E. (1989). *The evolving role of statistical assessments as evidence in the courts*. Springer: New York.

French, P., and Harrison, P. (2007). Position Statement concerning use of impressionistic likelihood terms in forensic speaker comparison cases. *International Journal of Speech Language and the Law*, *14*(1), 137–144.

French, P., Nolan, F., Foulkes, P., Harrison, P., and McDougall, K. (2010). The UK position statement on forensic speaker comparison: A rejoinder to Rose and Morrison. *International Journal of Speech Language and the Law*, *17*(1), 143–152.

Giles, H., Coupland, J., and Coupland, N. (1991). *Contexts of accommodation: Developments in applied sociolinguistics* (Cambridge University Press). New York.

Gonzalez-Rodriguez, J., Rose, P., Ramos, D., Toledano, D. T., and Ortega-Garcia, J. (2007). Emulating DNA: Rigorous quantification of evidential weight in transparent and testable forensic speaker recognition. *IEEE Transactions on Audio, Speech, and Language Processing*, *15*(7), 2104–2115.

Greenberg, C. S., Martin, A. F., Brandschain, L., Campbell, J. P., Cieri, C., Doddington, G. R., and Godfrey, J. J. (2010). Human assisted speaker recognition in NIST SRE10. *Odyssey 2010 – The Speaker and Language Recognition Workshop*.

Greenberg, C. S., Martin, A. F., Doddington, G. R., and Godfrey, J. J. (2011). Including human expertise in speaker recognition systems: report on a pilot evaluation. In *IEEE International Conference on Acoustics, Speech and Signal Processing (ICASSP), 5896–5899.*

Hautamäki, R. G., Hautamäki, V., Rajan, P., and Kinnunen, T. (2013). Merging human and automatic system decisions to improve speaker recognition performance. In *Proceedings of Interspeech*, 2519–2523.

Jessen, M. (2008). Forensic phonetics. *Language and Linguistics Compass*, 2(4), 671–711.

Kahneman, D. (2011). *Thinking, fast and slow.* Macmillan.

Kahn, J. (2011). *Parole de locuteur: performance et confiance en identification biométrique vocale.* Université d'Avignon.

Kahn, J.; Audibert, N.; Rossato, S.; Bonastre, J.-F. (2011). Speaker verification by inexperienced and experienced listeners vs. speaker verification system. In *IEEE International Conference on Acoustics, Speech and Signal Processing (ICASSP), 5912 –5915.*

Kahn, J., Audibert, N., Rossato, S., and Bonastre, J.-F. (2010). Intra-speaker variability effects of speaker verification performance. *Odyssey 2010 – The Speaker and Language Recognition Workshop.*

Kassin, S. M., Dror, I. E., and Kukucka, J. (2013). The forensic confirmation bias: Problems, perspectives, and proposed solutions. *Journal of Applied Research in Memory and Cognition*, 2(1), 42–52.

Kenny, P., Boulianne, G., Ouellet, P., and Dumouchel, P. (2007). Joint factor analysis versus eigenchannels in speaker recognition. *IEEE Transactions on Audio, Speech, and Language Processing*, *15*(4), 1435–1447.

Kenny, P., Gupta, V., Stafylakis, T., Ouellet, P., and Alam, J. (2014). Deep neural networks for extracting baum-welch statistics for speaker recognition. *Odyssey 2014 – The Speaker and Language Recognition Workshop.*

Kersta, L. G. (1962). Voiceprint identification. *Nature*, *196*(4861), 1253–1257.

Kim, M., Horton, W. S., and Bradlow, A. R. (2011). Phonetic convergence in spontaneous conversations as a function of interlocutor language distance. *Laboratory Phonology*, 2(1), 125–156.

Krauss, R. M., and Pardo, J. S. (2006). Speaker perception and social behavior: bridging social psychology and speech science. In P.A.M. van

Lange (ed.) Bridging *social psychology: Benefits of transdisciplinary approaches*, pp. 273–278.

Labov, W. (1972). *Sociolinguistic patterns* (University of Pennsylvania Press). Philadelphia.

Lelong, A., and Bailly, G. (2012). Characterizing phonetic convergence with speaker recognition techniques. *The Listening Talker Workshop (LISTA 2012)*, 28–31.

Lei, Y., Scheffer, N., Ferrer, L., and McLaren, M. (2014). A novel scheme for speaker recognition using a phonetically-aware deep neural network. In *IEEE International Conference on Acoustics, Speech and Signal Processing (ICASSP)*, 1695–1699.

Levitan, R., and Hirschberg, J. B. (2011). Measuring acoustic-prosodic entrainment with respect to multiple levels and dimensions. *Proceedings of Interspeech*, 3081–3084.

Mandasari, M. I., McLaren, M., and van Leeuwen, D. A. (2011). Evaluation of i-vector speaker recognition systems for forensic application. *Proceedings of Interspeech*, 21–24.

Martin, A., Doddington, G., Kamm, T., Ordowski, M., and Przybocki, M. (1997). The DET curve in assessment of detection task performance. *Proceedings of Eurospeech*, Rhodes.

Martin, A. F., Greenberg, C. S., Stanford, V. M., Howard, J. M., Doddington, J. J., and Godfrey, J. J. (2014). Performance factor analysis for the 2012 NIST speaker recognition evaluation. *Proceedings of Interspeech*. Singapore.

Matrouf, D., Bonastre, J.-F., and Fredouille, C. (2006). Effect of speech transformation on impostor acceptance. In *IEEE International Conference on Acoustics, Speech and Signal Processing (ICASSP)*, 933–936.

McLaren, M. L., Matrouf, D., Vogt, R. J., and Bonastre, J.-F. (2008). Combining continuous progressive model adaptation and factor analysis for speaker verification. *Proceedings of Interspeech*, Brisbane.

McLaren, M., Matrouf, D., Vogt, R., and Bonastre, J.-F. (2011). Applying SVMs and weight-based factor analysis to unsupervised adaptation for speaker verification. *Computer Speech & Language*, *25*(2), 327–340.

Morrison, G. S. (2011). Measuring the validity and reliability of forensic likelihood-ratio systems. *Science & Justice*, *51*(3), 91–98.

Morrison G. S. (2014). Distinguishing between forensic science and forensic pseudoscience: Testing of validity and reliability, and approaches to forensic voice comparison. *Science and Justice, 54*(3), 245–256.

Nakasone, H., and Beck, S. B. (2001). Forensic automatic speaker recognition. *Odyssey 2001 – The Speaker and Language Recognition Workshop.*

NIST. (2010). The NIST year 2010 speaker recognition evaluation plan. Available at http://itl.nist.gov/iad/mig/tests/sre/2010/NIST_SRE10_evalplan.r6.pdf

Ortega-Garcia, J., Gonzalez-Rodriguez, J., and Marrero-Aguiar, V. (2000). AHUMADA: A large speech corpus in Spanish for speaker characterization and identification. *Speech Communication, 31*(2–3), 255–264.

Pardo, J. S. (2006). On phonetic convergence during conversational interaction. *The Journal of the Acoustical Society of America, 119*(4), 2382–2393.

Pardo, J. S. (2013). Measuring phonetic convergence in speech production. *Frontiers in Psychology, 4*, 559.

Pardo, J. S., Gibbons, R., Suppes, A., and Krauss, R. M. (2012). Phonetic convergence in college roommates. *Journal of Phonetics, 40*(1), 190–197.

Perrot, P., Aversano, G., and Chollet, G. (2007). Voice disguise and automatic detection: Review and perspectives. In Y. Stylianou, M. Faundez-Zanuy, and A. Esposito (ed.), *Progress in nonlinear speech processing* (pp. 101–117). Springer Berlin Heidelberg.

Petrovska-Delacrétaz, D., Chollet, G., and Dorizzi, B. (2009). *Guide to biometric reference systems and performance evaluation.* Springer: London.

Phillips, P. J., Martin, A., Wilson, C. L., and Przybocki, M. (2000). An introduction evaluating biometric systems. *Computer, 33*(2), 56–63.

Pruzansky, S. (1963). Pattern-matching procedure for automatic talker recognition. *The Journal of the Acoustical Society of America, 35*(3), 354–358.

Przybocki, M., and Martin, A. F. (2004). NIST speaker recognition evaluation chronicles. *Odyssey 2004 – The Speaker and Language Recognition Workshop.*

Przybocki, M., Martin, A. F., and Le, A. N. (2006). NIST speaker recognition evaluation chronicles-part 2. *Odyssey 2006 – The Speaker and Language Recognition Workshop.*

Przybocki, M. A., Martin, A. F., and Le, A. N. (2007). NIST speaker recognition evaluations utilizing the mixer corpora – 2004, 2005, 2006. *IEEE Transactions on Audio, Speech and Language Processing*, *15*(7), 1951–1959.

Ramos, D., Franco-Pedroso, J., and Gonzalez-Rodriguez, J. (2011). Calibration and weight of the evidence by human listeners. The ATVS-UAM submission to NIST HUMAN-aided speaker recognition 2010. In *IEEE International Conference on Acoustics, Speech and Signal Processing (ICASSP)*, 5908–5911.

Reynolds, D. A., Quatieri, T. F., and Dunn, R. B. (2000). Speaker verification using adapted Gaussian Mixture Models. *Digital Signal Processing*, *10*(1–3), 19–41.

Roberts, P. (2013). Renegotiating forensic cultures: Between law, science and criminal justice. *Studies in History and Philosophy of Science Part C: Studies in History and Philosophy of Biological and Biomedical Sciences*, *44*(1), 47–59.

Rose, P. (2006). Technical forensic speaker recognition: Evaluation, types and testing of evidence. *Computer Speech & Language*, *20*(2–3), 159–191.

Rose, P., and Morrison, G. (2009). A response to the UK position statement on forensic speaker comparison. *The International Journal of Speech, Language and the Law*, *16*(1), 139.

Scherer, K. (1986). Vocal affect expression: A review and a model for future research. *Psychological Bulletin*, *99*, 143–165.

Schwartz, R., Campbell, J. P., Shen, W., Sturim, D. E., Campbell, W. M., Richardson, F. S., Dunn, R. B., and Granville, R. (2010). USSS-MITLL 2010 human assisted speaker recognition. In *IEEE International Conference on Acoustics, Speech and Signal Processing (ICASSP)*, 5904–5907.

Shen, W., Campbell, J.P., Straub, D., and Schwartz, R. (2011). Assessing the speaker recognition performance of naïve listeners using mechanical turk. In *IEEE International Conference on Acoustics, Speech and Signal Processing (ICASSP)*, 5916–5919.

Soong, F. K., Rosenberg, A. E., Juang, B.-H., and Rabiner, L. R. (1987). Report: A vector quantization approach to speaker recognition. *AT&T Technical Journal*, *66*(2), 14–26.

Stafylakis, T., Kenny, P., Senoussaoui, M., and Dumouchel, P. (2012). Preliminary investigation of boltzmann machine classifiers for speaker recognition. *Odyssey 20012 – The Speaker and Language Recognition Workshop.*

Thompson, W. C., Kaasa, S. O., and Peterson, T. (2013). Do jurors give appropriate weight to forensic identification evidence? *Journal of Empirical Legal Studies, 10*(2), 359–397.

Tosi, O. (1979). *Voice identification theory and legal applications.* Baltimore: University Park Press.

Tversky, A., and Kahneman, D. (1974). Judgment under uncertainty: Heuristics and biases. *Science, 185*(4157), 1124–1131.

Univaso, P., Soler, M. M., and Gurlekian, J. A. (2013). Human assisted speaker recognition using forced alignments on HMM. *International Journal of Engineering Research and Technology*, 2(9), ESRSA Publications.

van Dijk, M., Orr, R., van der Vloed, D., and van Leeuwen, D. (2013). A human benchmark for automatic speaker recognition. In *Proceedings of the 1st International Conference Biometric Technologies in Forensic Science, Nijmegen*, 39–45.

van Leeuwen, D. A., and Brümmer, N. (2013). The distribution of calibrated likelihood-ratios in speaker recognition. *Proceedings of Interspeech, Lyon.*

van Leeuwen, D. A., Martin, A. F., Przybocki, M. A., and Bouten, J. S. (2006). NIST and NFI-TNO evaluations of automatic speaker recognition. *Computer Speech &Language, 20*(2–3), 128–158.

Vasilakakis V., Cumani S., Laface P. (2013). Speaker recognition by means of Deep Belief Networks. *Proceedings of Biometric Technologies in Forensic Science, Nijmegen.*

Wu, Z., Siong, C. E., and Li, H. (2012). Detecting converted speech and natural speech for anti-spoofing attack in speaker recognition. *Proceedings of Interspeech, Portland.*

Zhang, C., and Tan, T. (2008). Voice disguise and automatic speaker recognition. *Forensic Science International, 175*(2–3), 118–122.

Speech Production and Perception

Edited by Susanne Fuchs and Pascal Perrier

Vol. 1 Susanne Fuchs / Melanie Weirich / Daniel Pape / Pascal Perrier (eds.): Speech Planning and Dynamics. 2012.

Vol. 2 Anne Hermes: Articulatory Coordination and Syllable Structure in Italian. 2013.

Vol. 3 Susanne Fuchs / Daniel Pape / Caterina Petrone / Pascal Perrier (eds.): Individual Differences in Speech Production and Perception. 2015.

www.peterlang.com

www.ingramcontent.com/pod-product-compliance
Ingram Content Group UK Ltd.
Pitfield, Milton Keynes, MK11 3LW, UK
UKHW041824031026
15587UKWH00002B/548

9 783631 665060